SCIENCE IN LIBERAL ARTS
COLLEGES

A Longitudinal Study of 49
Selective Colleges

W. Rodman Snelling
and
Robert F. Boruch

PUBLISHED FOR RESEARCH CORPORATION BY
COLUMBIA UNIVERSITY PRESS, NEW YORK AND LONDON,
1972

Copyright © 1972 Research Corporation
Library of Congress Catalog Card Number: 76-164502
ISBN: 0-231-03599-3
Printed in the United States of America

FOREWORD

Research Corporation's interest in the liberal arts colleges goes back to 1945 when it inaugurated its first formal grants program in the physical sciences. In this and succeeding programs of grants particular attention was given to the smaller institutions and those having limited interest in or financial resources for research.

In the late 1950s the foundation tested the concept of providing broad departmental support for the sciences in liberal arts colleges and smaller universities. The special characteristics of these institutions and the environment they afforded for academic research as well as for undergraduate education of scientists were regarded as especially promising.

That experiment evolved into Research Corporation's largest grants activity from 1960 through 1968. During this period the foundation provided support totaling some $6 million in the form of departmental grants for improvement of the sciences in nearly 100 predominantly liberal arts institutions. This program was phased out in 1968 when it was judged to have accomplished its basic purpose and as other and larger funding agencies began to make similar institutional grants to the colleges in support of their science programs.

BACKGROUND TO THE STUDY

While Research Corporation was concentrating its efforts on the college science program, there evolved considerable discussion and controversy as to the scientific productivity of the liberal arts colleges which in the first half of the century had been recognized as the major producers of science graduates going on to advanced work.

On the one hand, it was asserted that the undergraduate colleges which were not parts of universities were no longer attractive to young science teacher-scholars, or to highly qualified and motivated students; that the colleges' curricula and facilities were not keeping pace with those of the major institutions; and that they were becoming less able to provide competent candidates for the graduate institutions. On the other hand, it was argued that such conclusions were based on opinion or limited information, not on solid data, and that meaningful comparisons could not be made because information was lacking on the products of the institutions, particularly of the colleges of generally acknowledged quality.

Research Corporation decided that it might make a useful contribution by supporting an independent study to provide some pertinent information on science in liberal arts colleges. No attempt would be made to compare colleges one with another or with other types of institutions, or to encompass in the study all liberal arts colleges or even a selection possibly representative of the group. Rather, the study would examine intensively a well-defined sample of relatively high-quality institutions, probing for the factors which seemed to have some influence on productivity in the sciences.

The research design that was developed and the study which subsequently was conducted involved some 20,000 science and mathematics majors who had graduated over a ten-year period from a selected group of 49 liberal arts colleges. The data sought on the graduates included pre-college characteristics (the input), impact of the college (the "black box") and post-college activities (the output). The study results presented here should be of interest to all who are concerned with the motivation, education and utilization of scientists.

It should be noted, however, that the text has been written primarily for administrators and faculty members, rather than educational researchers, although the latter undoubtedly will find the tabular data of value.

A complete list of acknowledgments of those contributing to this study would take many pages, but I should like to single out Mr. Charles H. Schauer, Executive Vice President of Research Corporation, who sensed the need and timeliness of the study, conceived the overall plan for its conduct, and supervised the undertaking for the foundation. Research Corporation wishes to record its particular appreciation to the nine college presidents who constituted an informal advisory panel on the objectives of the study and who strongly supported its development, and also to the presidents, administrative officers, science faculties and graduates at the participating colleges, whose cooperation and assistance made the study possible. It is our hope that their large investment of time and effort will be justified by the findings presented in the following pages.

James S. Coles, President
Research Corporation

January 1971

CONTENTS

x

TABLES

Science in Liberal Arts Colleges

1

Introduction and Summary

The Research Corporation study of liberal arts science was undertaken to provide data relating to the productivity of a selected group of men's, women's and coeducational colleges. The study represents an in-depth examination of graduates of 49 institutions of overall high academic quality and at least moderate science orientation. It is a sample which should not be considered representative of all liberal arts colleges nor of any other group of institutions of higher education. Appendix A lists the participating colleges and describes the selection process.

W. Rodman Snelling, a statistician and educational consultant, was retained by Research Corporation in 1967 to develop the research design and to conduct the survey. The services of National Computer Systems, a Minneapolis processing center, were engaged for electronic scanning of completed questionnaires. Computer facilities were arranged with Florida Atlantic University in Boca Raton under the direction of William M. Hunt. Following compilation of the data in 1969, Robert F. Boruch, then at the American Council on Education, was retained as a consultant to work with Dr. Snelling on the analyses and to write the research report. This report is the joint effort of Drs. Snelling and Boruch.[1]

Each of the participating colleges has been furnished a com-

---

[1] Other publications relating to this work include: W. Rodman Snelling, "The Impact of a Personalized Mail Questionnaire," *The Journal of Educational Research,* 63, 3 (November 1969). W. Rodman Snelling and Robert F. Boruch, "Factors Influencing Student Choice of College and Course of Study," *Journal of Chemical Education,* 47 (May 1970), 326. W. Rodman Snelling, Robert F. Boruch, and Nancy B. Boruch, "Science Graduates of Private and Selective Liberal Arts Colleges," *College and University,* 46, 3 (Spring 1971).

puter print-out of its own data, together with a handbook for their interpretation. Particular care was taken to eliminate from the individual college print-outs certain categories of responses in order that confidential information would be respected; some of the data had been provided by graduates and members of faculties and administrations only after assurance had been given that anonymity would be preserved.

RESEARCH DESIGN

The findings presented here relate to science majors who graduated in the decade 1958–1967 from the 30 coeducational institutions, 10 men's colleges and 9 women's colleges which participated in the study. The study is limited to consideration of biology, physics, chemistry, mathematics and pre-medical undergraduate programs. Data have been obtained under a research paradigm which called for obtaining biographic attributes of entering students, institutional characteristics and the professional and extraprofessional accomplishments of graduates.

The research design included requirements for admissions and registrar data from the participating institutions, and questionnaire survey responses from department chairmen, faculty members and graduates. Five questionnaires were designed to elicit the data for the study. Appendix B contains samples of the questionnaires.

Admissions offices were asked for entrance data on prospective science majors, such as science and math grades and honors, probable major at entrance, type of high school attended, scholarship assistance at entrance and Scholastic Aptitude Test (S.A.T.) scores.

Registrars' offices were asked for information on science graduates, such as size of college class, rank in class, graduate record examination, grade-point averages and college honors earned.

The questionnaire for the science faculty requested information on the rank, age and service of the faculty member and his opinions on a number of items including esteem in which liberal arts science faculty is held, involvement of undergraduates in research, and strength of undergraduate science programs in liberal arts colleges as compared with university programs.

The questionnaire for department chairmen asked for the college policy as to the semester in which the student declares his major and the chairman's opinion as to this timing, the percentage of science majors who changed directions, the reasons for the changes, and factors motivating students to complete science majors, as well as a number of other matters of opinion.

The graduate questionnaire asked for the grade level at which a science major was selected, various factors which may have influenced that selection, financial contributions to graduate expenses, graduate school experience, and occupational and other information on postgraduate achievements.

Responses on admissions and registrar data were requested of the 49 participating colleges, and questionnaire survey responses asked of 206 department chairmen and 1,119 faculty members. Usable data on students were received from all 49 institutions, with the information being incomplete in the case of four colleges. Opinions on liberal arts science education were furnished by 157 department chairmen and 765 faculty members in 44 of these institutions. A six-page questionnaire mailed to 20,833 graduates of the period 1958–1967 was completed and returned by 16,395 graduates of all 49 participating colleges. The graduate sample comprised 11,543 men and 4852 women; of these 7993 men and 3152 women were from co-ed colleges. The high percentage of participation in the study leaves little doubt about the validity of the data obtained.

The research design sought an optimum return of 90% so that the conclusions drawn could not be challenged statistically. Within the test study of four colleges a return of 92% was achieved from the graduates. While the main study which provided the data con-

TABLE 1.1

Participating Colleges—Total 1967 Enrollment
and College Type

	Number of Colleges		
Total Enrollment	Co-ed	All-Male	All-Female
200–499	0	0	1
500–999	2	5	4
1000–2499	25	5	4
2500–4999	3	0	0

tained in this report did not equal the percentage achieved in the
test response, a review of the data indicates no substantial variations. It can therefore be safely assumed that the data of the main
study effectively describe the population.

Characteristics of the institutional sample are furnished in
Tables 1.1 and 1.2. The majority of colleges are small, having total
student enrollments in the range of 1000 to 2499. Most of the institutions in this particular modal class are coeducational colleges
(Table 1.1).

TABLE 1.2

Participating Colleges—Selectivity
and College Type

	Number of Colleges		
NMSQT Average	Co-ed	All-Male*	All-Female
113–120	5	0	0
121–128	14	3	4
128+	11	6	5

*Note: Selectivity data unavailable for one college.

"Selectivity" of the institutions can be characterized by average grades of entering freshmen on national, normative tests of academic performance, such as Scholastic Aptitude Tests and National Merit Scholarship Qualifying Tests (NMSQT). Table 1.2 shows that all of the institutions of this sample have freshmen who fall into the top three categories of an achievement index, based on average NMSQT scores for each college, and the modal category is the maximum achievement possible (128+).[2] This particular group of colleges recruited high achievers at a higher rate than did a majority of liberal arts colleges in this country.

With respect to financial affluence, nearly 85% of the institutions are in the highest bracket, with these institutions receiving approximately $2,500 per student per year in 1967. The data available from Creager and Sell show that the national median for all institutions at that time was in the range of $1,750–$2,000 in total revenues per student per year.

In this sample group, slightly more than one-third of the science graduates had majored in biology, 22% were graduated in mathematics, 20% in chemistry, 12% in physics, and the remaining individuals had graduated from pre-medical programs. Substantially more than one-half the group were men.

SUMMARY OF FINDINGS

The major part of the study data following is summarized in the form of percentages. The total of returns was sufficiently close to the population to make tests of statistical significance of little value.

---

[2] The tabulated data have been drawn from a standardized information file based on attributes of more than 2,000 colleges and universities. The information system furnishes complete coverage of a wide variety of institutional characteristics. Creation and development of the data is available from J. A. Creager and C. L. Sell, "The Institutional Domain of Higher Education: A Characteristics File for Research," *ACE Research Reports,* Washington, D.C., 4, 6 (1969).

Religious, Ethnic and Racial Backgrounds. The sample group is remarkably homogeneous with respect to religious, ethnic and racial composition. Nearly two-thirds of the graduates are Protestant, and about 60% are at least third-generation American in each of the graduating classes from 1958 to 1967. Catholics and Jews together account for about 17% of the entire group and the ratio has increased slightly over the period considered. Racial group representation is consistent with the preceding information: more than 90% of the graduates are white, and the ratio of Negroes and Oriental students, though small, has increased. The stability of the ratios when comparing across graduating classes is marked, suggesting the consistency of selection of colleges by students and the admissions programs of the institutions.

The most heterogeneous institutional group appears to be men who graduate from coeducational institutions. They are most variable with respect to ethnic origins and the "age" of their families. The largest numbers of male Negro and Oriental students graduated from coeducational institutions.

Precollege Achievement. High school achievements and precollege performance level, as reflected by Scholastic Aptitude Test scores, are uniformly high for the group. Average scores for most recent graduates are in the upper 10% of national achievement levels on quantitative tests. Graduates are in the upper fifth of performance on verbal tests. Sex differentials are apparent insofar as graduates of all-male colleges achieve higher quantitative score averages and women's averages for verbal scores are above the men's. Trends over the period 1958–1967 are obvious, with average scores increasing more than 50 points overall. Increases in S.A.T. scores are more evident in the quantitative area. Comparisons across major discipline categories on the verbal and (especially) the quantitative scores demonstrate differential interests and previous training of the groups. Increases in score averages and decreases in score variability over time can be attributed to strengthening of precollege training and increasingly stringent admissions policies.

Decisions on Science, Majors and Advanced Degrees. The age at which various decisions are made to specialize constitute useful information for college counseling agencies. Between 50 and 65% of all graduates in this study indicated that science was chosen as their major field of study during their elementary school years, or during the first year of secondary education. Moreover, the recent graduates acknowledged early decisions more frequently. The increments in ratio for recent graduates may generally be attributable to societal pressures for an early decision, to earlier maturation, and to primary and secondary school counseling.

As one might expect, the student's choice of a *specific* major field of study is delayed relative to the time acknowledged for determining a general field of endeavor. More than half the graduates said that a decision to enter a specialized area of science was made during the college years. Only about 5% made their choice during the first nine years of education. Women appear to make a decision at a later stage in their development than men.

Similar conclusions may be drawn from the data on advanced degree aspirations. Trends toward earlier decisions are evident, although the majority of students appear to settle on a specific objective during their last year of college. Women defer making a choice more frequently than men.

Determinants in Choice of College. Major influences in determination of college choice are, of course, important in vocational counseling and for college student recruitment efforts. From 60 to 75% of the science graduates reported that small classes coupled with close faculty-student ties and small total enrollment were primary considerations in their decision to enter a liberal arts college. The range in percentages is a function of the sex of the student and college type. A smaller, but still notable, ratio designated flexibility of curriculum, quality of students and adequacy of facilities as major influences. Sex differences are apparent insofar as women are more likely to indicate major parental impact (15–20%) than are the men (10%). On the other hand, many men consider "assured admission" and availability of scholarship aid important in college selection.

Undergraduate Financial Support. The support of the student's education can be expected to be heavily dependent on parental income levels. In this sample, the more recent graduates provide evidence of a high affluence level for the group. Nearly one-fifth of respondents said family income was $25,000 or more annually. Less than 15% designated a bracket of less than $13,000 per year. A majority of respondents received more than half their required expenses directly through parents. Women affirm nearly complete financial support more frequently than the men do.

Reliance on loans has increased during the period under consideration but typically accounts for less than one-third of complete support. Men are more likely to rely on loans than are women.

Some form of scholarship aid was evident for approximately 45% of the graduates. About 15% of the 1967 graduates received support for at least half their educational expenses through scholarships. Comparing across disciplines, graduates of pre-med and biology curricula are least likely to receive support through scholarships.

The largest ratio of working students occurs in coeducational institutions. More than one-third of them had earned 20–30% of necessary expenses during the school year. Men from the co-ed colleges were more likely to have worked during the summer also.

Transfers to and from Science Departments. Department chairmen from biology, chemistry, physics and mathematics departments within the liberal arts colleges included in this study supplied their perceptions on frequency of and causes for attrition from their departments.

Results of this survey indicate that physics and mathematics students who transfer are most likely to transfer out of their major discipline and biology students are least likely to do so. Of the biology and mathematics students who transfer, it is found that the majority choose nonscience departments, while physics and chemistry students are as likely to choose another science major as they are to choose a nonscience major if a transfer is decided upon. Roughly 5 to 10% of students who transfer from the college science departments enter universities. Chairmen have indicated, however, that the majority of transfers occur within the college itself. These observations are uniform across all departments. Reasons for attrition are found to be largely a function of the particular department. The proposal that laboratory time commitments are overburdening is found to be only a minor influence on the student's decision to transfer, but is acknowledged more frequently to be a major influence by chemistry and physics department chairmen. Chairmen from all departments except biology consider the lack of mathematical competence to be a major influence on students who transfer. Inability to comprehend advanced theories is a major deterrent in all departments, but especially in mathematics. Biology students are perceived to transfer largely because of the greater personal challenge offered by other fields.

Undergraduate Achievement. Grade-point averages have a predictable distribution, the modal level comprising B−, C+, and B. Female graduates of both co-ed and women's colleges achieve higher performance levels somewhat more frequently than men. Differentials appear at the A+ level and A− to B+ level, where 2% and 5%, respectively, more women than men attain these averages. Considering the major disciplines, grades are more heterogeneous for math and physics groups. Pre-med and biology graduates attain high grades less frequently, and the modal grade levels have higher ratios.

Phi Beta Kappa awards were made at higher rates to graduates of the men's colleges, with 12 to 18% of the 1958–1967 students receiving this recognition.

Graduate School Attendance and Achievement. More than three-fourths of the earliest graduating classes (1958–1963) had enrolled in graduate training. Advanced studies were acknowledged less frequently by recent graduates but the ratio is still notable. The highest proportion of doctorates are awarded to male graduates of co-ed or men's colleges. The men's college graduates are most likely to obtain a medical degree, relative to the other groups. Law degrees are acquired occasionally, but far more students obtain advanced degrees in their specialty. Data on exact field of specialization in graduate school suggest some major shifts in graduates' interests during their late college years and early in the course of their advanced training. Respondents whose undergraduate education emphasized biology or mathematics shift to other advanced disciplinary areas at a higher rate than physics, chemistry, and pre-med majors.

Financial Support for Graduate School. From 70 to 80% of men who have enrolled in graduate schools have received some form of stipend during their course of studies. The medium of support may be research or teaching assistantships and work-free stipends for tuition alone or tuition plus expenses. The most evident increases during 1958–1967 have been in the category of work-free support. The ratio of male and female graduates receiving this form of support has increased. Sources of stipends are diverse, but the graduate college itself provides more than one-fourth of students with direct financial or employment assistance. As one would expect, the National Science Foundation is a major source of support to the extent that frequency of stipends and grants is surpassed only by institutional allocations. Private foundations and National Defense and Education Act monies furnish additional students with aid at a rate which is only a bit lower than these other sources.

Performance levels in graduate schools, as reflected by grades, are typically somewhat higher than achievements in undergraduate school. Much of the difference, of course, is attributable to policies of the graduate departments and to selection of high quality students. Sex differentials in grades appear but are not as strong as in the undergraduate data, suggesting more homogeneity of student abilities and motivation.

Postgraduate Employment. Because of graduate school commitments, full-time employment is delayed for many individuals in this sample. The majority of male respondents defer entrance to the job market until four years after graduation. Women differ insofar as they are more likely to obtain full-time employment soon after graduation, and they are more frequently unemployed three to five years following graduation.

Business and commercial enterprises are the single largest employer group for men. Both business and secondary or primary school work attract a high ratio of women. College and university teaching, research or administration account for slightly more than one-tenth of male graduates.

Occupations within these major employee classifications differ widely depending on the sex of the student and major discipline. Biology undergraduate majors are found primarily in teaching jobs, laboratory or hygiene work. More than 15% obtain employment as college teachers, dentists, or engineering technicians. Data on chemistry majors suggest that they will probably become chemists after graduation, but more than 15% of the group enter medicine, laboratory and technical support work in engineering science. Physics undergraduate majors contribute strongly to engineering manpower pools. More than one in four of these students enter college teaching in physics or work as physicists after graduation and advanced studies. Computer programming attracts many graduates in this and the mathematics field.

Postgraduate Professional Achievement. Postgraduate achievement and activities are impressive for extent and diversity. Small percentages of men had written books; the ratios decrease from 2% of the earliest classes to nearly zero for more recent groups. The analogous ratios for women are depressed below the men's level slightly and these contributions are made a bit later than the men's. Published articles are more evident for both men and women. Nearly 15% of the older graduates have written for journals or the popular media. Inventing attracts fewer men and women but the proportion of men who have achieved in such activity is notably 2%. Research grant acquisition is, of course, related to the other professional endeavors and the relations are reflected in the percentage data also. Although the rate of recent male graduates who have received awards is low, nearly one-tenth of graduates during 1958–1967 received awards. For women, the statistics are small in magnitude and variable from year to year.

Professional memberships and honorary awards are important aspects of the graduate's development. Between 25 and 35% of each graduating class comprises individuals with membership in one professional organization. From early graduating classes, nearly equal ratios occur for membership in more than one organization.

Postgraduate Extraprofessional Activities. Information on reading habits, civic and political involvement, and other organized activities was solicited from the graduates of the colleges considered in this study.

The majority of the graduates indicated that daily newspapers are read either frequently or constantly. News magazines and

books not related to professional expertise are read somewhat less frequently, but do constitute a considerable portion of the graduates' reading material. Literary magazines receive little attention from the graduates. The majority of men and women do use a library card after graduation with men more frequently making use of this service. In later years a larger percentage of men use a library card while a smaller percentage of women do so.

Participation by men and women in civic organizations is limited to less than one-fifth of the total number of respondents in each group and has decreased over the years. Slightly more attention is devoted to educational organizations with women more likely to be involved than men. Women are also more likely to be involved in political groups.

Memberships in fraternities (or sororities) is characteristic of a very small percentage of the respondents (10%). Service groups, such as the Red Cross, also receive a relatively small amount of attention. Participation in church or synagogue activities is typical of a minority of respondents, with women slightly more apt to be involved than men. Club membership activities are found to be fairly attractive to a majority of the graduates. A substantial proportion of the respondents are participants on community service committees.

Only a small percentage of the respondents have held leadership positions in civic, educational, or service groups; an even smaller ratio have been seated in local political office.

Attendance at cultural events is typical of roughly one-half the graduates and this proportion remains fairly constant over time.

Opinions of Graduates and Department Chairmen. Graduates and department chairmen were asked their opinions on several aspects of the undergraduate liberal arts education.

Department chairmen were asked to give their opinions on curricular requirements and orientation of the undergraduate education offered at their colleges. Distribution requirements of the undergraduate curricula were not considered to have a negative impact on a student's choice of his particular major. A majority of department chairmen indicated that preparation for graduate school is the principal concern governing administrative and educational emphasis. Involvement through research or project work was considered a major orientation by fewer than 20% of the chairmen.

Science graduates of liberal arts colleges generally hold their institutions in high esteem. Opinions on laboratory training, advanced coursework, and especially faculty quality are markedly positive. The majority of the graduates have indicated, however, that most of the attributes considered should be strengthened by increased emphasis. For the most part, all graduates felt that their colleges had provided them with excellent training, and the vast majority indicated that, given a second chance, they would reselect a liberal arts college for pursuit of their education. Indeed, when asked to evaluate their institution relative to other types of institutions and other liberal arts colleges, the overwhelming majority rated their own institution as excellent.

A COMMENT ON THE DATA

This study is intended to document only the science and mathematics graduates of liberal arts institutions during the period 1958 to 1967. It is evident that the sample of graduates under examination is unique, characterized by very high abilities and diverse background factors and interests. Insofar as much of the data show stability with respect to time, then inferences (by way of extrapolation) about more recent students can be made safely. However, current major changes in college environments which bear on the data collected here should be considered in developing judgments about current liberal arts students. Because the sample is deliberately limited to private liberal arts colleges, generalizations to

technical institutions or to universities and public institutions must be qualified with other data on the character of these groups. Some gross differences between these types of institutions and between attributes of graduates from the institution can be made by utilizing contemporary census-type reports on the college population.

2

Religious, Ethnic and
Racial Backgrounds

The religious, ethnic and racial backgrounds of the graduates reflect the nature of student population attracted to the high-prestige liberal arts college. Graduates are remarkably homogeneous in these characteristics and this consistency is maintained through the 1958–1967 period. The respondents are mainly Caucasian and Protestant. They come from families whose national origins are England or other West European countries. Substantial numbers of the graduates indicate that they are at least third-generation American.

	1958 N = 1236 %	1959 N = 1292 %	1960 N = 1419 %	1961 N = 1589 %
Invalid	1.5	1.4	0.9	1.4
No response	2.8	2.6	2.9	2.1
Agnostic	6.6	7.6	8.9	7.6
Atheist	2.6	1.6	3.0	2.8
Catholic	10.5	8.9	8.1	7.7
Jewish	6.6	7.0	7.9	8.2
Protestant	67.7	67.9	66.2	68.0
Other	1.5	3.1	2.0	2.2

Religious background data on science and mathematics majors are provided in Tables 2.1, 2.2 and 2.3. Approximately two-thirds of all graduates responding are Protestant. Catholics and Jews each account for 8–9% of the entire group. Agnostics comprise an additional 8%, whereas only a small percentage (2–3%) have atheistic backgrounds. Although these statistics are nearly constant over the ten-year period (Table 2.1), slight trends are discernible. Protestant representation, for example, decreased about 5%, while Jewish representation increased by about the same amount. Representation in the various religious categories is nearly independent of sex and field of study (Tables 2.2 and 2.3). There is a small indication, however, that graduates whose parents were atheist or agnostic were most probably enrolled in physics, while Jewish offspring were more likely found in pre-medicine, biology, or chemistry, in that order. As shown in Table 2.3, the largest percentage of agnostics are men in co-ed colleges, while the smallest percentage of Jewish students are women in co-ed colleges.

TABLE 2.1

Religious Origins of Graduates,
by Year of Graduation

1962 N = 1591	1963 N = 1724	1964 N = 1800	1965 N = 1919	1966 N = 1854	1967 N = 1942
%	%	%	%	%	%
1.2	1.3	1.1	1.3	1.2	0.8
3.1	2.8	2.0	2.6	1.3	1.2
8.4	6.2	7.3	8.2	8.7	8.4
1.8	3.3	2.7	2.6	3.1	3.2
8.4	8.1	9.1	8.3	9.8	9.9
8.1	8.9	9.7	8.8	9.1	11.1
66.6	67.6	66.1	66.0	64.2	62.9
2.6	1.9	1.9	2.5	2.6	2.3

TABLE 2.2

Religious Origins of Graduates,
by Major Subject

	Biology N = 5996 %	Chemistry N = 3362 %	Math N = 3666 %	Physics N = 2045 %	Pre-med N = 1329 %
Invalid	1.3	1.1	1.0	1.7	1.0
No response	2.3	2.3	2.0	3.0	2.0
Agnostic	7.3	8.0	7.9	10.5	5.6
Atheist	2.6	2.7	2.5	3.8	1.8
Catholic	8.9	9.0	9.3	8.8	7.5
Jewish	10.4	8.5	6.6	4.9	13.4
Protestant	65.2	66.1	68.5	64.3	67.4
Other	2.1	2.4	2.4	3.1	1.3

	1958 N = 1236 %	1959 N = 1292 %	1960 N = 1419 %	1961 N = 1589 %
Invalid	15.9	16.3	17.2	16.1
No response	3.5	3.3	3.2	2.6
African	0.7	0.5	0.2	0.4
American Indian	0.0	0.0	0.1	0.1
Arab	0.2	0.4	0.1	0.1
East European	7.8	8.3	9.0	9.4
English	37.9	34.4	32.0	33.8
Indian	0.1	0.1	0.0	0.1
Mediterranean	1.8	2.1	2.1	1.3
Nordic	5.8	4.3	5.2	5.5
Oriental	0.5	1.3	1.9	0.9
Slavic	1.9	1.9	2.0	2.1
South American	0.1	0.2	0.1	0.0
West European	24.0	27.2	27.1	27.6

TABLE 2.3

Religious Origins of Graduates,
by College Type

	Male Co-ed N = 7993 %	All-Male N = 3550 %	Female Co-ed N = 3152 %	All-Female N = 1700 %
Invalid	1.1	0.9	1.3	1.9
No response	2.3	2.6	1.8	2.5
Agnostic	9.6	5.6	6.4	6.6
Atheist	3.2	2.1	2.2	2.6
Catholic	8.3	10.3	7.6	11.3
Jewish	8.9	11.0	3.1	13.3
Protestant	64.0	65.9	75.5	59.7
Other	2.7	1.6	2.1	2.1

TABLE 2.4

Ethnic Origins of Graduates,
by Year of Graduation

1962 N = 1591 %	1963 N = 1724 %	1964 N = 1800 %	1965 N = 1919 %	1966 N = 1854 %	1967 N = 1942 %
17.7	16.7	17.6	17.3	16.6	16.9
3.2	2.8	2.3	2.6	1.7	1.1
0.8	0.6	0.9	0.8	1.0	1.1
0.1	0.1	0.0	0.1	0.1	0.1
0.1	0.3	0.1	0.1	0.2	0.1
8.9	9.6	10.1	10.8	9.6	11.0
31.1	31.8	30.8	31.3	31.8	28.3
0.2	0.2	0.1	0.2	0.1	0.0
1.7	1.4	1.9	1.6	1.8	1.4
4.7	5.5	5.2	4.9	6.0	5.4
0.9	1.1	0.9	0.9	1.2	1.3
2.8	2.7	2.1	2.8	2.2	2.8
0.0	0.0	0.1	0.1	0.1	0.2
27.9	27.3	26.8	26.8	27.6	30.3

TABLE 2.5

Ethnic Origins of Graduates,
by College Type*

	Male Co-ed N = 7993	All-Male N = 3550	Female Co-ed N = 3152	All-Female N = 1700
	%	%	%	%
Invalid	16.3	12.8	21.4	19.7
No response	2.6	2.8	2.1	2.5
African	0.8	0.8	0.5	0.7
American Indian	0.1	0.1	0.0	0.0
Arab	0.2	0.1	0.0	0.2
East European	10.1	11.0	5.5	11.8
English	28.8	39.7	31.5	32.4
Indian	0.1	0.1	0.1	0.1
Mediterranean	1.9	1.9	0.8	1.8
Nordic	6.4	3.6	5.2	3.4
Oriental	1.1	0.6	0.9	2.5
Slavic	2.8	2.0	2.0	2.7
South American	0.1	0.1	0.0	0.0
West European	28.6	24.6	30.0	22.5

*Note: The high invalid response rate is attributed to a weakness in the question. Most graduates presumably used ethnic origin of their fathers if origins of parents were different.

Ethnic origins are presented in Tables 2.4, 2.5 and 2.6. Data on this attribute are complicated by graduates' multiple responses on the questionnaire. A number of invalid responses are attributable to interethnic marriages: different origins for mother and father, or grandparents. Over the ten years considered, representation of each ethnic group was fairly constant. The smallest percentage of students had American Indian, Arabian, Indian and South American backgrounds, while students with English and

West European heritage claim the major percentages. The male graduates of co-ed colleges appear to be the most heterogeneous with respect to ethnic origin (Table 2.5). Men's colleges, on the other hand, are the most homogeneous group. Representation of various ethnic groups does not differ substantially across major curriculum categories (Table 2.6). The percentages within each discipline approximate those found within the college-type category.

The majority of graduates' families have been in the United

TABLE 2.6

Ethnic Origins of Graduates,
by Major Subject

	Biology N = 5996 %	Chemistry N = 3362 %	Math N = 3666 %	Physics N = 2045 %	Pre-med N = 1329 %
Invalid	18.2	17.8	15.1	15.4	15.6
No response	2.5	2.6	2.4	3.0	2.3
African	.6	.8	.8	.6	.9
American Indian	.1	.0	.1	.1	.1
Arab	.1	.2	.1	.2	.2
East European	10.6	9.8	8.2	7.1	11.9
English	32.6	28.7	32.6	31.8	36.7
Indian	.1	.2	.1	.1	.1
Mediterranean	1.8	1.9	1.6	1.6	1.2
Nordic	4.7	5.4	5.8	6.7	3.8
Oriental	.9	1.7	.9	1.4	.8
Slavic	2.3	2.6	2.6	2.3	2.9
South American	.0	.0	.1	.2	.0
West European	25.5	28.4	29.7	29.8	23.6

	1958 N = 1236 %	*1959* N = 1292 %	*1960* N = 1419 %	*1961* N = 1589 %
American generation				
Invalid	0.8	0.5	1.3	0.8
No response	5.5	5.0	5.1	5.5
First generation	4.9	6.0	5.9	5.2
Second generation	10.9	8.4	12.0	11.5
Third generation	17.4	19.6	18.2	18.7
More than third	60.5	60.5	57.4	58.3
Racial heritage				
Invalid	0.2	0.1	0.0	0.0
No response	6.4	5.6	7.7	6.9
Oriental	0.8	1.2	2.2	1.0
Negro	0.6	0.4	0.4	0.3
White	92.1	92.7	89.7	91.8

TABLE 2.8

American Generation and Racial Heritage of Graduates,
by College Type

	Male Co-ed N = 7993 %	*All-Male* N = 3550 %	*Female Co-ed* N = 3152 %	*All-Female* N = 1700 %
American generation				
Invalid	0.6	0.3	1.2	1.4
No response	5.1	5.1	4.0	5.4
First generation	4.6	3.2	3.7	6.4
Second generation	11.2	10.4	10.3	12.3
Third generation	21.8	18.0	18.1	18.9
More than third	56.7	63.0	62.9	55.7
Racial heritage				
Invalid	0.1	0.0	0.1	0.1
No response	7.3	7.0	4.5	5.6
Oriental	1.3	0.7	1.0	2.7
Negro	0.5	0.3	0.4	0.8
White	91.0	91.9	94.0	90.9

TABLE 2.7

American Generation and Racial Heritage of Graduates,
by Year of Graduation

1962 N = 1591 %	1963 N = 1724 %	1964 N = 1800 %	1965 N = 1919 %	1966 N = 1854 %	1967 N = 1942 %
0.8	0.9	0.6	0.8	0.4	0.4
4.9	3.8	3.3	3.4	3.6	3.5
4.8	3.8	3.3	3.4	3.6	3.5
10.6	11.7	11.7	10.2	10.6	11.6
19.8	18.3	22.6	21.6	20.3	21.2
59.2	60.3	57.8	58.8	60.7	58.8
0.1	0.1	0.1	0.1	0.0	0.0
7.0	6.8	5.9	7.0	5.8	6.0
1.2	1.1	1.1	1.2	1.3	1.4
0.3	0.3	0.4	0.4	0.6	0.8
91.4	91.8	92.5	91.3	92.3	91.8

TABLE 2.9

American Generation and Racial Heritage of Graduates,
by Major Subject

	Biology N = 5996 %	Chemistry N = 3362 %	Math N = 3666 %	Physics N = 2045 %	Pre-Med N = 1329 %
American generation					
Invalid	0.8	0.9	0.7	0.5	0.6
No response	5.3	4.7	4.5	5.3	4.1
First generation	3.8	5.7	3.8	5.1	3.6
Second generation	11.4	11.4	10.3	9.9	11.1
Third generation	20.5	20.5	18.9	20.1	18.9
More than third	58.3	56.8	62.0	59.1	61.6
Racial heritage					
Invalid	0.0	0.1	0.1	0.0	0.1
No response	6.7	6.6	5.7	7.1	6.9
Oriental	1.0	1.6	1.1	1.8	1.0
Negro	0.5	0.7	0.4	0.3	0.3
White	91.8	91.0	92.9	90.8	91.7

States for at least three generations (Tables 2.7, 2.8 and 2.9). This aspect of graduate background remains remarkably stable over the decade—about 60% of the respondents are at least third generation in each graduating class. No real increases in the representation of "younger" families are evident across college types, but males in co-ed colleges are somewhat more heterogeneous, having fewer graduates from the higher generation category. The stability of these data is a reflection of the students' criteria for selection of a college; it may also be indicative of admissions policies which include parental attributes (i.e., sons and daughters of alumni receiving preferential treatment), or the student's college choice may reflect parental preference.

Representation of various racial groups in the sample is markedly stable throughout the decade (Table 2.7). Somewhat more than 90% of respondents are white, and the percentage of Negroes and Orientals increases only slightly to reach 0.8% and 1.4%, respectively, in 1967. Within college type (Table 2.8), the largest percentage of Negroes and Orientals are women graduates of all-female colleges. Women's colleges appear to be somewhat more heterogeneous than other college types with respect to racial composition: 2.7% Oriental and 0.8% Negro. Respondents from Oriental families have majored predominantly in chemistry and physics, and Negroes are best represented in chemistry (Table 2.9).

The paucity of Negro graduates within the major disciplines considered here is suggestive. However, inferences must be qualified by current knowledge about choice of discipline by Negro students. The current social and legal emphasis on expansion of opportunity for Negro students is limited by the self-selection of these students to specific disciplines. If admissions policies include emphasis on increases in racial mix of the student groups within the sciences, then special efforts to encourage entry into the sciences must be made. Active recruitment and financial support of such students in science curricula may be warranted in the physical sciences as well as in the social sciences.

3

Scholastic Aptitude Test Scores

Scholastic Aptitude Test (S.A.T.) scores are a convenient index for assessing students' verbal and quantitative skills. The tests, issued by the College Entrance Examination Board, are meaningful (albeit imperfect) predictors of a student's abilities and achievements, and are used in a large number of institutions as part of admissions criteria. In test creation, a deliberate attempt is made to ensure comparability of test scores from one year to the next. That is, a score of 350 is roughly indicative of the same level of ability in 1958 as it is in 1967. Test scores range from 200 to 800, and the annual median test score is approximately 500. A detailed analysis of the verbal and quantitative aptitude scores of the respondents in this study is presented in this section.

In the current study, both verbal and quantitative S.A.T. score averages increase markedly through 1958–1967, regardless of the college type, major field of study, or sex of the individuals tested. The increments are strongly indicative of the increased abilities of high school students who enroll at the institutions, and the increasing selectivity of the colleges during this period. Evidence of increasing uniformity in application of the test scores in admissions criteria is indicated by the diminishing standard deviation of scores. The steady decrease in this index of student variability also reflects a gradual increase in the average level of student abilities as measured by the tests. There may also be a self-selection factor operating. That is, fewer students with relatively low abilities may be applying to the colleges during this ten-year period, and this limitation in the pool of available students enhances the average of scores for entering freshmen and restricts

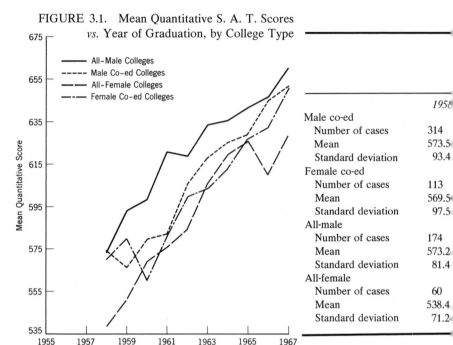

FIGURE 3.1.   Mean Quantitative S. A. T. Scores *vs.* Year of Graduation, by College Type

Legend:
- All-Male Colleges
- Male Co-ed Colleges
- All-Female Colleges
- Female Co-ed Colleges

	1958
Male co-ed	
Number of cases	314
Mean	573.5
Standard deviation	93.4
Female co-ed	
Number of cases	113
Mean	569.5
Standard deviation	97.5
All-male	
Number of cases	174
Mean	573.2
Standard deviation	81.4
All-female	
Number of cases	60
Mean	538.4
Standard deviation	71.2

the range of these scores. Quantitative score averages for each college type, gender, and major discipline are presented as a function of year of graduation in Tables 3.1 through 3.4. These data are then plotted in Figures 3.1 through 3.4.

Graduates of men's colleges obtained higher quantitative scores than the other groups, and scores showed a smaller deviation over the ten-year period. Male and female graduates of the co-ed colleges obtained lower scores generally, and the score variability is slightly greater. Lowest quantitative averages are found for the

TABLE 3.1

Mean Quantitative S.A.T. Scores and Standard Deviations
of Entering Freshmen,
by Year of Graduation and by College Type

1959	1960	1961	1962	1963	1964	1965	1966	1967
369	509	652	736	850	945	1002	868	945
566.03	579.46	581.90	605.47	617.83	624.94	626.88	644.89	651.27
93.0	98.15	95.09	90.70	91.12	84.91	83.06	74.49	78.78
130	196	259	266	362	326	358	355	369
579.61	559.91	580.82	599.60	603.12	612.56	626.59	632.19	650.47
95.04	95.26	90.67	86.85	81.02	76.52	75.41	76.02	72.88
243	235	292	300	323	370	353	345	356
593.35	598.30	620.43	618.50	633.46	635.36	641.16	646.25	660.52
86.23	89.27	89.54	88.65	80.01	77.32	76.11	71.08	71.62
69	51	70	66	81	78	108	86	92
551.28	568.90	575.46	584.02	605.67	619.26	625.38	609.70	628.09
88.59	79.24	84.06	76.23	78.81	68.28	65.40	79.49	79.21

graduates of women's colleges (Figure 3.1, Table 3.1). The lower averages for women's colleges on the quantitative test are attributed to the type of curriculum offered as well as high school student ability levels. Emphasis on liberal arts curricula in the colleges may tend to attract students with a greater verbal than quantitative expertise.

Verbal S.A.T. score averages (Figure 3.2 and Table 3.2) are higher for the female graduates; women from co-ed institutions obtained scores which were generally lower than those for gradu-

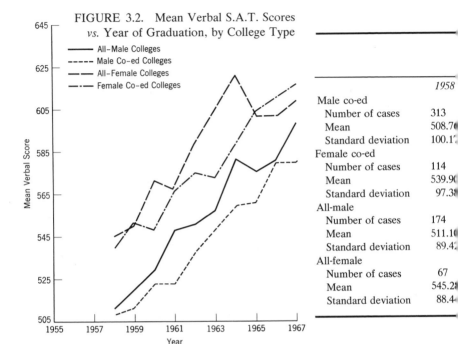

FIGURE 3.2.   Mean Verbal S.A.T. Scores vs. Year of Graduation, by College Type

— All-Male Colleges
---- Male Co-ed Colleges
— — All-Female Colleges
—·— Female Co-ed Colleges

1958

Male co-ed	
Number of cases	313
Mean	508.7
Standard deviation	100.1
Female co-ed	
Number of cases	114
Mean	539.9
Standard deviation	97.3
All-male	
Number of cases	174
Mean	511.1
Standard deviation	89.4
All-female	
Number of cases	67
Mean	545.2
Standard deviation	88.4

ates of women's colleges. Again, the graduates of the male colleges were more homogeneous at the time of entrance to college than other groups. Scores achieved by these men are slightly, but consistently, higher than those achieved by male graduates of co-ed colleges.

The rate of increase in both verbal and quantitative scores over the ten years considered is approximately the same whether one considers graduates of single-sex or co-ed colleges.

Trends within various major subject areas for quantitative

TABLE 3.2

Mean Verbal S.A.T. Scores and Standard Deviations of Entering Freshmen,
by Year of Graduation and by College Type

1959	1960	1961	1962	1963	1964	1965	1966	1967
370	509	652	737	852	948	1003	865	946
512.41	523.16	522.92	537.87	548.06	559.63	560.59	579.18	578.97
101.48	103.20	89.70	93.95	89.45	85.91	84.27	81.64	84.09
130	195	260	266	362	328	358	356	369
551.91	547.87	565.80	574.76	572.42	589.66	603.08	610.33	616.00
103.74	99.59	87.78	89.71	88.89	76.14	73.94	85.83	78.25
243	235	293	301	323	370	352	346	356
520.80	529.41	548.20	550.72	557.42	581.23	575.37	580.68	598.23
87.45	93.65	85.72	91.86	83.79	81.03	73.28	76.77	71.82
64	52	71	69	81	79	108	86	92
550.36	571.29	566.70	589.54	604.05	620.32	601.61	602.38	609.45
98.61	85.25	84.78	83.41	82.89	70.75	72.87	89.26	78.47

S.A.T. scores are illustrated in Figure 3.3 and Table 3.3. Mathematics and physics majors achieved higher scores, on the average, than did respondents from the other disciplines. A slight differential rate of increase in the direction of higher achievement by mathematics majors relative to physics majors is evident. The trend may be attributed to increasing reliance on quantitative admissions tests for selection of students in math, and increased reliance on such tests by students themselves in choosing their major field of study. Chemistry, biology, and pre-med graduates' quantitative score averages are arrayed in that order. Consistently lower scores by pre-med and biology majors than by any of the other students are evident.

In the verbal tests (Figure 3.4, Table 3.4), differences across subject areas are not so striking as in the case of quantitative scores. Data on students from the earlier graduating classes suggest that chemistry and physics majors have had higher scores

FIGURE 3.3.   Mean Quantitative S.A.T. Scores *vs.* Year of Graduation, by Major Subject

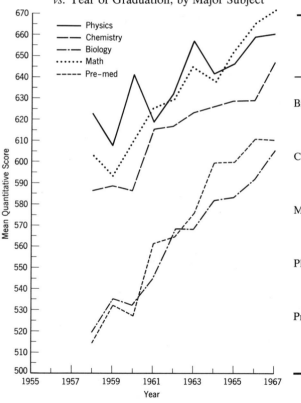

	1958
Biology	
Number of cases	155
Mean	520.5
Standard deviation	84.8
Chemistry	
Number of cases	97
Mean	585.8
Standard deviation	70.9
Mathematics	
Number of cases	79
Mean	601.7
Standard deviation	71.6
Physics	
Number of cases	87
Mean	619.9
Standard deviation	81.5
Pre-medicine	
Number of cases	29
Mean	516.5
Standard deviation	78.7

in the verbal tests. More recent test results suggest nearly equivalent verbal skills for graduates within all disciplines. The premed verbal mean is somewhat lower than the mean found for other disciplines. A trend in the direction of increased homogeneity in verbal skills is apparent. The standard deviations, within nearly all groups, decrease slightly across time.

The generally increasing aptitude scores are suggestive of both

TABLE 3.3

Mean Quantitative S.A.T. Scores and Standard Deviations
of Entering Freshmen,
by Year of Graduation and by Major Subject

1959	1960	1961	1962	1963	1964	1965	1966	1967
174	209	266	257	279	248	255	226	205
534.95	531.72	546.00	567.49	569.79	581.88	583.53	592.18	610.00
89.22	84.98	85.48	79.95	82.46	78.29	76.37	69.49	74.72
122	142	167	163	153	123	100	61	67
587.58	585.32	614.25	617.39	621.27	625.09	628.41	628.72	650.96
83.63	91.25	90.80	93.46	71.73	75.13	67.44	76.03	73.64
92	134	189	234	225	239	227	181	152
591.58	607.68	623.87	630.63	644.55	638.64	656.89	668.31	673.72
91.08	86.66	83.62	82.31	74.80	69.94	62.93	60.34	60.33
67	117	117	131	101	87	76	61	43
606.81	638.72	616.09	635.76	656.07	637.33	648.14	658.11	660.09
66.48	80.43	90.57	79.19	71.17	81.45	72.01	64.78	74.19
22	24	32	26	23	24	11	8	19
534.73	527.71	561.59	568.54	575.87	599.25	599.91	610.00	610.26
92.38	95.66	69.64	70.97	74.66	85.38	69.51	95.04	83.91

FIGURE 3.4.   Mean Verbal S.A.T. Scores
vs. Year of Graduation, by Major Subject

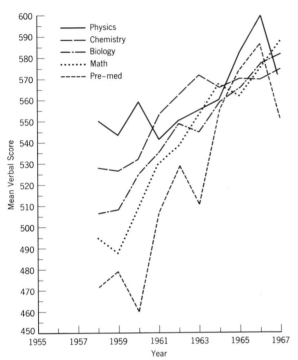

	1958
Biology	
Number of cases	158
Mean	506.3
Standard deviation	90.0
Chemistry	
Number of cases	99
Mean	527.8
Standard deviation	87.7
Mathematics	
Number of cases	79
Mean	494.3
Standard deviation	93.8
Physics	
Number of cases	88
Mean	548.7
Standard deviation	95.4
Pre-medicine	
Number of cases	29
Mean	471.2
Standard deviation	89.0

the competitiveness among students for entrance to the sample colleges, and the increasing educational sophistication of the high school seniors taking such tests. More importantly, increment in quality and extent of postgraduate performance is likely to be associated with the rise in level of ability. Evidence for the latter presumption is examined in the following sections of this study.

A compelling feature of these data is the apparent increasing

TABLE 3.4

Mean Verbal S.A.T. Scores and Standard Deviations
of Entering Freshmen,
by Year of Graduation and by Major Subject

1959	1960	1961	1962	1963	1964	1965	1966	1967
174	209	268	258	279	250	255	226	205
507.20	524.30	534.68	548.74	541.69	557.66	566.39	578.49	579.00
93.65	91.29	88.12	84.53	83.75	77.61	81.43	76.57	76.16
123	143	168	163	153	124	100	61	67
526.14	531.96	552.71	560.55	570.21	563.43	567.71	569.77	574.88
99.60	103.00	95.90	99.96	89.96	79.82	81.59	83.15	89.94
92	133	189	234	226	242	227	181	152
487.53	507.53	530.37	539.45	554.73	566.20	560.70	576.79	588.70
91.45	105.32	93.07	93.39	91.74	83.48	85.67	84.33	84.36
67	117	116	131	102	88	76	61	43
544.48	559.61	540.24	548.06	571.58	557.99	577.42	596.75	568.98
97.54	99.38	98.85	87.36	87.32	82.94	77.51	74.19	81.49
22	24	32	26	23	24	11	8	19
477.41	459.50	504.97	527.77	510.04	556.92	574.09	586.88	549.32
74.15	84.54	79.09	86.42	77.34	79.23	63.20	119.97	106.82

emphasis on aptitude test scores in selection of students. Admissions criteria are obviously meritocratic in this regard. If such criteria are heavily weighted, the students' entrance to the college will be largely a function of the technical skills acquired in the secondary schools. Insofar as reliance on such tests increases, the correspondence of the particular secondary school environment to the probability of being accepted by the college increases.

4

Decisions on Science Majors and Advanced Degrees

This section contains a description of the temporal aspects of the graduates' decisions to enter the sciences and a specific major field of study. Respondents also provided information about the time at which they decided upon the highest degree that they would seek.

The data are important insofar as career and curriculum gui-

	1958 N = 1236 %	1959 N = 1292 %	1960 N = 1419 %	1961 N = 1589 %
No response	0.6	0.6	0.9	1.0
Prior to Grade 9	37.7	36.8	43.6	42.0
Grade 9	13.2	14.4	14.1	14.4
Grade 10	19.4	19.2	16.0	18.5
Grade 11	13.1	11.9	10.3	11.2
Grade 12	5.3	6.0	7.0	5.5
College				
First year	5.7	6.4	4.4	4.3
Second year	4.1	3.6	2.6	2.3
Third year	0.5	0.9	0.4	0.5
Fourth year	0.2	0.2	0.1	0.1
Fifth year	0.1	0.1	0.1	0.0
Sixth year	0.0	0.0	0.0	0.0
Invalid	0.2	0.0	0.5	0.1

dance may have a differential impact on the student's development, depending on when such counseling is provided. That is, if decisions about major emphasis on science studies are made early, then later general (rather than specific) counseling may be unnecessary and ineffectual. Prediction of enrollments in the various science departments and longer-range manpower studies can be facilitated using these data, at least for the type of institutions which we have considered.

CHOICE OF SCIENCE AS A FIELD

Interestingly, from 37 to 51% of graduates acknowledged that a decision to make science their major field of interest was made before the ninth grade (Table 4.1). The age of choice is probably

TABLE 4.1

Grade Level at Which Science Was Chosen
As Major Field of Interest,
by Year of Graduation

1962 N = 1591 %	1963 N = 1724 %	1964 N = 1800 %	1965 N = 1919 %	1966 N = 1854 %	1967 N = 1942 %
0.5	0.8	0.6	0.5	0.5	0.2
42.4	45.0	50.7	47.8	48.8	51.2
17.7	16.8	14.7	15.8	15.2	14.8
15.8	16.5	16.2	16.7	16.2	14.8
11.0	10.3	8.2	8.7	9.2	9.0
5.9	4.4	4.6	4.5	4.1	4.5
4.1	3.3	2.8	3.3	3.5	3.0
2.3	2.0	1.5	2.2	2.0	2.1
0.3	0.6	0.5	0.5	0.2	0.4
0.0	0.1	0.1	0.0	0.0	0.0
0.1	0.1	0.0	0.0	0.1	0.1
0.0	0.0	0.1	0.0	0.1	0.0
0.1	0.2	0.1	0.1	0.3	0.1

a function of parental education, as well as the nature of the educational courses encountered before the choice.

From the longitudinal data of Table 4.1, it is evident that earlier decisions on major study area are made by an increasingly larger percentage of students as time goes on. The rise is likely to be a function of the increasing emphasis on counseling in the high schools during recent years, and perhaps also the increasing pressures on the student who is ambivalent about the choice of his career. In any event, the importance of grade school years in shaping the choice of a career for the science majors appears to be primary.

Nearly half the chemistry, physics, and pre-med students decided to enter the sciences during their first eight years of edu-

TABLE 4.2

Grade Level at Which Science Was Chosen
As Major Field of Interest,
by Major Subject

	Biology N = 5996 %	Chemistry N = 3362 %	Math N = 3666 %	Physics N = 2045 %	Pre-med N = 1329 %
No response	0.8	0.5	0.3	0.5	0.9
Prior to Grade 9	44.3	47.3	42.1	50.4	49.3
Grade 9	13.9	15.1	18.5	13.7	14.2
Grade 10	20.2	13.5	16.2	13.5	14.2
Grade 11	7.8	12.7	11.3	11.7	8.3
Grade 12	4.6	6.1	5.2	5.4	3.5
College					
First year	5.1	2.9	3.5	3.2	3.6
Second year	3.4	1.5	2.2	1.4	2.0
Third year	0.6	0.2	0.5	0.1	1.2
Fourth year	0.1	0.0	0.1	0.0	0.3
Fifth year	0.0	0.1	0.1	0.0	0.2
Sixth year	0.0	0.0	0.0	0.0	0.0
Invalid	0.2	0.1	0.1	0.1	0.2

TABLE 4.3

Grade Level at Which Science Was Chosen
As Major Field of Interest,
by College Type

	Male Co-ed N = 7993 %	All-Male N = 3550 %	Female Co-ed N = 3152 %	All-Female N = 1700 %
No response	0.6	0.8	0.5	0.6
Prior to Grade 9	48.2	49.6	37.9	35.7
Grade 9	15.0	13.6	17.5	15.3
Grade 10	14.9	17.2	20.0	18.7
Grade 11	10.0	9.1	10.7	11.6
Grade 12	4.8	4.2	5.5	7.2
College				
First year	3.3	2.9	4.7	7.5
Second year	2.2	2.1	2.6	2.9
Third year	0.6	0.3	0.4	0.3
Fourth year	0.1	0.1	0.0	0.0
Fifth year	0.1	0.1	0.0	0.0
Sixth year	0.0	0.0	0.0	0.0
Invalid	0.1	0.1	0.2	0.2

cation, while 42% of mathematics and 44% of biology majors made
their choices during this same period (Table 4.2). Evidence of
the influence of high school coursework is apparent since 20% of
those in biology chose science at the tenth-grade level, a traditional year for introduction of biology in high schools.

Considering all disciplines, less than 10% of the graduates indicated that a scientific field of interest was adopted during their
college years. More biologists made their choices during this period than did graduates in other disciplines. Physicists appeared
to be the earliest to make a choice, with only about 5% having
decided upon science during four or more years of college.

Sex differences do not appear to have a substantial effect on the time at which a student chooses to enter the field of science (Table 4.3). However, more women (graduates of co-ed or women's colleges) made their decision later than men. The largest differential occurs between the graduates of the women's and the men's colleges; 10% of the women and 6% of the men acknowledged a decision during their college years. Men and women from co-ed colleges are more homogeneous insofar as 6% and 7.5%, respectively, made this decision while in college.

CHOICE OF MAJOR

Although the choice of science as a field of study occurs rather early, picking a specific major area was delayed for a majority

	1958 N = 1236 %	1959 N = 1292 %	1960 N = 1419 %	1961 N = 1589 %
No response	1.2	0.9	1.6	0.9
Prior to Grade 9	4.9	4.0	6.3	5.2
Grade 9	2.4	2.5	2.4	2.1
Grade 10	4.4	5.7	4.9	6.5
Grade 11	8.1	8.8	8.7	9.1
Grade 12	15.9	16.6	16.1	14.6
College				
First year	24.4	23.6	21.2	21.5
Second year	29.5	29.5	29.0	30.8
Third year	7.8	7.6	8.3	8.2
Fourth year	1.3	0.7	1.3	0.8
Fifth year	0.1	0.0	0.0	0.1
Sixth year	0.1	0.1	0.1	0.0
Invalid	0.0	0.2	0.1	0.3

of graduates (Table 4.4). About 58% made such a decision while in college, and only 4–6% acknowledged selection of their major before the ninth grade. No pronounced tendency to make earlier decisions is evident. Women graduates of the co-ed colleges and men from all-male and co-ed colleges are most alike since they selected major fields at approximately the same time (Table 4.5). Graduates of the all-female colleges delayed their preferences to some degree—about 64% choosing a major during their first or second year of college, as compared to about 50% of the other groups doing so. In all groups, the largest percentage chose majors in their sophomore year.

Across major field categories differences are not substantial. Pre-med students made their decisions earlier than students in other majors; 14% selected their major interest prior to the ninth

TABLE 4.4

					Grade Level at Which Major Was Selected, by Year of Graduation
1962 N = 1591 %	1963 N = 1724 %	1964 N = 1800 %	1965 N = 1919 %	1966 N = 1854 %	1967 N = 1942 %
0.8	1.2	0.9	0.6	1.0	0.4
5.0	5.5	5.8	4.9	4.9	5.5
2.6	2.5	2.9	3.2	2.3	2.4
5.9	5.8	6.9	7.1	8.4	8.1
10.8	10.2	10.5	10.4	10.0	10.7
15.8	14.9	15.6	14.1	15.6	14.5
21.9	22.6	23.8	23.8	23.1	22.2
28.8	29.6	26.0	27.7	27.7	28.3
7.7	6.4	6.4	7.1	6.1	7.0
0.4	0.9	0.7	0.9	0.8	0.7
0.1	0.2	0.1	0.1	0.0	0.1
0.1	0.0	0.0	0.1	0.0	0.1
0.1	0.2	0.3	0.1	0.2	0.1

TABLE 4.5

	Grade Level at Which Major Was Selected, by College Type			
	Male Co-ed *N = 7993* %	*All-Male* *N = 3550* %	*Female Co-ed* *N = 3152* %	*All-Female* *N = 1700* %
No response	0.9	1.0	0.9	0.7
Prior to Grade 9	5.9	5.3	4.6	2.9
Grade 9	2.5	2.5	3.1	1.9
Grade 10	6.4	6.3	7.5	5.7
Grade 11	9.9	10.0	10.6	7.8
Grade 12	16.2	14.5	15.6	12.2
College				
First year	21.5	21.1	23.8	30.8
Second year	27.1	30.7	27.5	33.2
Third year	8.1	7.5	5.9	4.7
Fourth year	1.2	0.8	0.4	0.2
Fifth year	0.1	0.0	0.1	0.0
Sixth year	0.0	0.1	0.0	0.0
Invalid	0.2	0.1	0.1	0.2

grade (Table 4.6). Somewhat less than 4% of physics, chemistry and mathematics graduates admitted that a selection had been made this early. The largest percentage of those choosing their major during the college years occurred in the mathematics group (65%). Nearly 10% fewer graduates in the other disciplines chose a major during this period.

DECISION TO SEEK HIGHEST DEGREE

A slight trend toward earlier development of degree aspirations is evidenced by the data of Table 4.7. The majority of the graduates elected to pursue a specific degree during their college years.

Within the college period, the fourth year was an important time for the majority of graduates in deciding on the highest degree to be sought.

Women generally postponed a choice of their ultimate degree until their fourth year of college more frequently than men did (Table 4.8). Female graduates of both women's colleges and the co-ed colleges are similar in this regard. Slightly more than one-fifth of the men acknowledged the fourth year as the decision point, and proportionately more men decided on the highest degree they would seek during the preceding years, including the grade school period.

Across major field categories, pre-med graduates acknowledged

TABLE 4.6

Grade Level at Which Major Was Selected,
by Major Subject

	Biology N = 5996 %	Chemistry N = 3362 %	Math N = 3666 %	Physics N = 2045 %	Pre-med N = 1329 %
No response	1.1	0.7	0.7	1.0	1.1
Prior to Grade 9	5.5	3.8	3.7	3.6	14.0
Grade 9	2.6	1.6	2.8	2.1	4.5
Grade 10	8.9	4.4	4.8	4.3	9.5
Grade 11	7.9	13.4	9.1	9.7	12.2
Grade 12	12.3	19.2	13.5	20.3	15.9
College					
First year	23.7	24.2	21.0	24.1	18.6
Second year	30.3	25.8	33.1	27.3	17.4
Third year	6.6	6.2	9.9	6.4	6.0
Fourth year	0.9	0.4	1.3	0.8	0.5
Fifth year	0.1	0.1	0.1	0.1	0.0
Sixth year	0.0	0.1	0.0	0.1	0.1
Invalid	0.2	0.1	0.1	0.2	0.2

	1958 N = 1236 %	1959 N = 1292 %	1960 N = 1419 %	1961 N = 1589 %
No response	6.0	4.6	5.8	6.4
Prior to Grade 9	6.3	6.4	8.0	7.6
Grade 9	2.4	2.2	1.8	1.6
Grade 10	2.5	2.6	3.0	2.4
Grade 11	2.4	3.2	3.0	2.7
Grade 12	5.6	4.9	6.5	5.2
College				
First year	6.1	6.7	4.3	6.8
Second year	7.2	6.3	6.0	6.5
Third year	11.8	10.8	11.2	12.2
Fourth year	27.0	27.7	27.5	27.6
Fifth year	10.0	12.4	12.4	10.6
Sixth year	12.7	12.2	10.4	10.5
Invalid	0.1	0.0	0.1	0.0

early decisions on degree aspirations more frequently than other groups (Table 4.9). The choice, of course, is associated with their decision to major in medicine. Mathematicians (11%) and physicists (15%) made decisions on highest degree to be sought before entering college less frequently than biologists or chemists.

In summary, nearly half the graduates decided during their primary school years to pursue interest in the sciences. Choice of a particular major area was late by comparison, occurring for most students while they were in college. Most graduates acknowl-

TABLE 4.7

1962 N = 1591 %	1963 N = 1724 %	1964 N = 1800 %	1965 N = 1919 %	1966 N = 1854 %	1967 N = 1942 %
			Grade Level at Which Highest Degree Sought Was Decided Upon, by Year of Graduation		
5.3	5.0	4.9	5.5	5.8	6.3
7.8	8.4	7.9	6.5	8.3	10.3
2.0	1.7	1.7	2.0	2.5	2.3
3.2	3.3	3.4	3.7	4.3	4.6
3.0	2.6	3.4	4.1	3.9	4.5
3.8	4.4	5.2	5.7	5.6	5.0
4.7	5.8	5.6	6.5	5.9	6.6
7.2	7.8	8.4	8.2	9.0	8.7
12.3	12.7	12.4	15.1	12.4	15.1
27.0	28.5	27.4	25.2	27.1	24.7
12.1	11.0	11.1	8.9	8.9	9.2
11.7	9.0	8.3	8.6	6.5	2.7
0.1	0.0	0.2	0.1	0.0	0.0

edged that a preference for highest degree to be sought was developed during their college years, and especially during their fourth year.

A definite trend toward an earlier decision to specialize in the sciences is apparent. There is a small tendency for more frequent early choices of major discipline and of degree aspirations as years pass.

Intergroup differences are not substantial, but graduates of women's colleges appear to delay most frequently decisions on

TABLE 4.8

	Male Co-ed N = 7993 %	All-Male N = 3550 %	Female Co-ed N = 3152 %	All-Female N = 1700 %
			Grade Level at Which Highest Degree Sought Was Decided Upon, by College Type	
No response	4.8	4.6	7.4	7.9
Prior to Grade 9	8.2	11.1	4.8	4.9
Grade 9	2.1	2.7	1.3	1.4
Grade 10	3.7	4.7	2.0	1.7
Grade 11	3.5	5.1	2.0	1.5
Grade 12	5.3	7.0	3.9	3.2
College				
First year	6.4	6.8	4.9	3.8
Second year	8.2	9.1	5.7	5.7
Third year	13.1	13.1	11.9	12.2
Fourth year	24.3	19.5	35.6	38.0
Fifth year	10.8	8.1	12.3	11.4
Sixth year	9.5	8.3	8.3	8.6
Invalid	0.1	0.1	0.0	0.0

science as a major field, a specific major, and the highest degree to be sought.

Across major field categories, physics and mathematics graduates appear to make choices somewhat later than other groups. Pre-med graduates most frequently affirm decisions of this sort at a younger age.

TABLE 4.9

	Biology N = 5996 %	Chemistry N = 3362 %	Math N = 3666 %	Physics N = 2045 %	Pre-med N = 1329 %
				Grade Level at Which Highest Degree Sought Was Decided Upon, by Major Subject	
No response	5.6	4.1	7.0	6.4	3.8
Prior to Grade 9	9.5	7.5	3.8	4.7	16.8
Grade 9	2.7	1.8	1.0	0.7	4.2
Grade 10	4.2	3.2	1.3	1.9	8.0
Grade 11	3.5	3.5	1.9	2.7	7.2
Grade 12	5.7	5.4	3.1	5.1	8.1
College					
First year	6.0	6.7	4.4	5.0	9.1
Second year	7.8	8.7	6.3	6.5	10.1
Third year	11.7	16.2	12.9	13.1	8.4
Fourth year	25.6	26.4	35.3	25.5	12.5
Fifth year	9.6	9.0	13.0	14.6	5.5
Sixth year	7.9	7.5	10.0	13.8	6.3
Invalid	0.0	0.1	0.1	0.0	0.2

5

Determinants in Choice of a College

One can hypothesize a wide variety of influences which affect a student's decision to attend a liberal arts institution. What attributes are most appealing to the students who choose this type of environment? Which features are acknowledged by graduates to have had no influence at all on their decision? These and other questions are considered in this chapter. The basic data on "strongly influential" factors in choosing a liberal arts college are presented in Tables 5.1 and 5.2.

One might expect that the institutional size, coupled with direct implications of this characteristic of liberal arts colleges, are important determinants in the choice of the college. Indeed, this is the case. Small classes, with associated close faculty-student ties, and a small total enrollment are acknowledged by a majority of graduates to be strong influences. The respondents also indicated that other functional attributes of the liberal arts college were important. The broad educational background and flexibility of curricula and the quality of students are only a bit less frequently cited as being important in choosing the college. From 45 to 55% of the students in each of the disciplines admitted the influential character of such features (Table 5.1).

Some of the factors that influenced a minority of graduates are interesting. The reputation of the college in mathematics and sciences was a strong influence for less than one-fourth of graduates in all major fields except pre-med. In the latter category about half the graduates provided affirmative responses. Impressions of the science facilities are important for fewer graduates;

TABLE 5.1

Major Influences Leading to a Decision
on Entering a Liberal Arts College,
by Major Subject

	Biology N = 5996 %	Chemistry N = 3362 %	Math N = 3666 %	Physics N = 2045 %	Pre-med N = 1329 %
...all student body	69.1	66.1	66.5	58.6	69.5
...exibility of liberal arts curriculum	45.1	40.3	45.9	38.6	39.1
...all classes and close faculty-student ties	73.9	71.5	65.4	63.3	74.1
...ctivities less competitive than at larger schools	11.2	9.4	10.5	8.5	14.2
...oad background offered by the liberal arts	54.5	49.6	51.9	48.9	50.5
...elatives attended the college	7.3	7.3	7.7	7.3	11.4
...ose friends were attending	3.8	3.4	3.1	3.6	6.8
...ollege was near home	7.1	8.8	10.3	11.6	9.7
...ollege was far from home	5.5	5.7	6.2	4.7	3.7
...holarship was granted	17.3	25.9	25.9	25.5	14.0
...ollege noted for strength in math and science	25.5	29.7	14.9	22.2	51.7
...dmission was assured	8.6	9.0	9.6	9.4	7.1
...pressed by college facilities	29.8	24.4	23.5	17.5	25.1
...pressed by science facilities	13.0	12.5	6.1	8.8	18.6
...omotion of alumni	6.6	6.5	5.7	6.0	8.9
...pposite sex close at hand	11.3	9.2	11.5	9.6	8.9
...ecommendation by school counselor	11.9	11.2	12.3	10.9	11.2
...ecommendation by high school science teacher	5.0	6.1	6.7	6.4	5.1
...rents wanted liberal arts college	13.1	11.1	12.7	11.8	11.7
...fluence of admissions staff	8.9	7.5	7.6	6.5	5.8
...uality of students in the college	53.4	45.7	46.6	41.5	54.7
...fluence of church	2.9	2.2	2.9	2.4	4.2

TABLE 5.2

Major Influences Leading to a Decision
Entering a Liberal Arts Colle
by College Ty

	Male Co-ed $N = 7993$ %	All-Male $N = 3550$ %	Female Co-ed $N = 3152$ %	All-Fem $N = 17$ %
Small student body	63.6	68.2	74.0	64.4
Flexibility of liberal arts curriculum	37.0	41.6	51.0	59.4
Small classes and close faculty-student ties	68.1	73.9	71.3	70.5
Activities less competitive than at larger schools	13.1	12.9	5.8	2.7
Broad background offered by the liberal arts	44.8	51.8	59.7	70.9
Relatives attended the college	8.2	7.3	7.6	6.8
Close friends were attending	4.3	4.3	3.0	1.8
College was near home	10.4	7.0	9.6	4.9
College was far from home	4.6	3.9	8.5	7.0
Scholarship was granted	22.2	21.8	21.1	20.9
College noted for strength in math and science	28.7	31.4	20.6	9.4
Admission was assured	9.6	8.8	8.7	6.1
Impressed by college facilities	20.0	25.9	29.3	42.1
Impressed by science facilities	11.9	13.6	9.2	7.8
Promotion of alumni	5.9	9.0	5.0	6.7
Opposite sex close at hand	11.8	0.8	20.5	6.5
Recommendation by school counselor	12.3	10.8	10.6	12.4
Recommendation by high school science teacher	6.2	5.4	6.1	4.5
Parents wanted liberal arts college	10.3	9.7	15.4	21.8
Influence of admissions staff	6.7	9.0	9.0	8.1
Quality of students in the college	40.2	52.1	55.0	72.1
Influence of church	2.5	2.7	4.8	0.4

less than 20% of the respondents indicated that this is a major influence.

Factors that influence relatively small but equal percentages of respondents include proximity of the college to the student's home, the fact that relatives had attended the college or that admission was assured, and the impact of the college admissions staff. From 5 to 10% of respondents affirmed the importance of these.

Of the standard types of guidance counseling available for choosing a college (high school counselor, science teacher and parents), the high school teacher was designated by the smallest percentage of graduates as having exerted a strong influence. Parents are only a bit more likely than counselors to have a strong impact.

Consider now the possibility that the major influences in choosing a liberal arts college interact with the various college types. It is plausible to expect influences to differ, depending on whether the respondent is a graduate of a co-ed, all-male, or all-female college. Table 5.2 provides information on this aspect of the college choice situation.

The women graduates more frequently cite the flexibility of curricula and broad background provided in the liberal arts college as strong influences in their choice. Women are more likely to be influenced by their parents than are the men in this sample. On the other hand, men acknowledged the impact of the college's reputation in math and science, and the science facilities more frequently. Also, men appear to be a bit more likely to be influenced by an assured college admission, scholarship aid and promotion of the college alumni.

Small classes, close faculty ties and a small student body are influential to about the same extent across all group categories.

Some differences between affirmative response rates associated with co-ed colleges and other groups are evident. Graduates of women's colleges indicated, more frequently than other groups, that impression of college facilities, parental advice, quality of the students at the college, and flexibility of the curriculum were important determinants of their college choice.

Respondents in this sample also provided information on factors which they perceived as deterrents in their choice. The presumption, of course, is that these negative aspects were not substantial enough to warrant choice of another type of college or attrition prior to graduation. In fact, few graduates admitted any negative impact of the features considered. The largest single category of response was the proximity of the college to home: between 5 and 7% said that the college was too close, and from 6 to 8% suggested that the distance was too far. The proportion of graduates acknowledging other deterring factors was negligible (much less than 3% for all the factors considered).

In summary, the strongly influential factors in choice of a liberal arts college are related to size of the student body and concurrent class size and faculty-student relations. A second grouping of major influence includes the flexibility of the curriculum offered and the broad background obtainable. Influence of traditional personal guidance (teachers, counselors, parents) forms a third ranking group with some discernible impact on student choice. According to the great majority of graduates, no single factor produced a substantial negative effect on their deliberation.

6

Undergraduate Financial Support

The adequacy of financial resources for students is a dominant theme in general theorizing about colleges and universities. This chapter contains answers to a variety of questions relevant to income sources of the graduates canvassed in this study. The basic framework of parental income levels is presented, together with an examination of the extent to which students relied on this source of income. Other means of support—loans, scholarships, part-time work, and personal savings—and the degree of dependency by students on these sources are explained.

FAMILY AID

Table 6.1 shows that parental income levels indicated by graduates generally increased over the ten-year period considered. Responses for the most recent classes (1967) provide substantial evidence of parental affluence. In 1967 19% of respondents indicated parents' income above $25,000 per year (national surveys of all institutions show only 7% in this bracket), and 13% indicated a familial income range of $19,000 to $24,999. A minority (43%) of students' parents fell into an income bracket of less than $13,000 per year. The rise in income during 1958–1967 is likely to be an inflationary effect. However, the rate of increase may be moderated by increasing enrollment of students from income brackets below the national median for the prescribed year. More detailed comparisons using national data need to be conducted in order to examine this hypothesis.

Parental income level, of course, strongly influences availabil-

	1958 N = 1236 %	1959 N = 1292 %	1960 N = 1419 %	1961 N = 1589 %
Invalid	0.0	0.0	0.1	0.0
No response	3.2	2.5	2.7	3.1
Under $7,000	25.5	24.9	22.7	20.4
$7,000–$12,999	34.2	37.2	35.6	35.7
$13,000–$18,999	15.3	14.6	17.7	17.9
$19,000–$24,999	9.2	7.7	8.7	9.8
Over $25,000	12.6	13.2	12.5	13.1

Percentage of Support	1958 N = 1236 %	1959 N = 1292 %	1960 N = 1419 %	1961 N = 1589 %
Invalid	0.5	0.1	0.5	0.3
No response	2.2	2.4	1.8	1.7
0%	4.1	5.0	4.5	6.0
10%	6.5	7.0	6.9	7.4
20%	6.3	5.7	5.2	4.9
30%	5.5	5.7	5.2	6.9
40%	6.0	5.3	5.9	5.5
50%	8.9	7.7	9.7	6.8
60%	7.4	7.4	6.3	8.1
70%	8.0	8.3	8.7	8.6
80%	11.3	12.0	12.3	11.6
90% and up	33.4	33.6	33.2	32.3

TABLE 6.1

Income Levels of Graduates' Families,
by Year of Graduation

1962 N = 1591 %	1963 N = 1724 %	1964 N = 1800 %	1965 N = 1919 %	1966 N = 1854 %	1967 N = 1942 %
0.1	0.0	0.1	0.1	0.1	0.1
3.1	2.4	2.2	2.9	3.5	3.1
19.2	16.7	14.2	14.3	12.2	12.0
35.3	35.3	36.5	34.8	33.6	31.1
19.0	20.9	21.0	21.4	23.3	22.2
9.4	11.1	10.2	10.6	12.3	12.9
14.0	13.7	15.7	16.0	15.1	18.7

TABLE 6.2

Percentage of Undergraduate Expenses Supported by Family,
by Year of Graduation

1962 N = 1519 %	1963 N = 1724 %	1964 N = 1800 %	1965 N = 1919 %	1966 N = 1854 %	1967 N = 1942 %
0.1	0.3	0.5	0.4	0.5	0.4
1.9	1.1	1.2	1.9	1.7	1.3
4.3	3.0	3.8	3.5	4.3	3.4
6.6	7.1	5.1	6.6	5.5	7.0
6.2	6.7	6.1	6.3	5.9	5.4
6.0	5.2	7.2	6.6	7.0	7.0
5.7	6.0	6.2	5.8	6.2	5.8
8.2	8.7	7.9	8.0	7.4	7.5
7.4	7.1	7.6	7.2	8.0	7.6
8.4	8.0	9.2	9.2	9.2	9.3
11.3	13.1	11.2	12.8	13.5	13.4
34.1	33.8	34.0	31.9	30.9	31.9

ity of funds for undergraduate expenses and the alternative sources of income utilized by students. Nearly 70% of the students said that 50% or more of their financial aid was contributed by parents (Table 6.2). The proportion is consistent over the years 1958–1967. Approximately one-third of the graduates received more than 90% of their assistance from families.

Table 6.3 contains the percentages of expenses supported by family categorized by sex and type of college. One would expect the women to be supported more frequently and to a greater extent by parents, and indeed this is the case. Women at female colleges most frequently affirm nearly complete support by their families. Over 52% of the respondents indicated at least 90% support by their families. Of female graduates of co-ed colleges, only 42% fell into this category. Men at co-ed colleges were less fre-

TABLE 6.3

Percentage of Undergraduate Expenses Supported by Family, by College Type

Percentage of Support	Male Co-ed N = 7993 %	All-Male N = 3550 %	Female Co-ed N = 3152 %	All-Female N = 1700 %
Invalid	0.4	0.4	0.3	0.4
No response	1.7	2.0	1.7	0.9
0%	5.4	3.9	2.6	1.4
10%	7.7	7.4	4.0	3.8
20%	6.7	6.4	3.9	4.4
30%	7.0	6.3	5.1	5.1
40%	6.4	5.8	5.3	4.8
50%	8.8	7.9	7.6	5.4
60%	8.2	6.6	7.2	5.9
70%	9.7	9.4	7.0	6.1
80%	12.8	12.1	12.7	9.7
90% and up	25.3	31.9	42.7	52.1

TABLE 6.4

Percentage of Undergraduate Expenses Supported by Family,
by Major Subject

Percentage of Support	Biology N = 5996 %	Chemistry N = 3362 %	Math N = 3666 %	Physics N = 2045 %	Pre-med N = 1329 %
Invalid	0.3	0.3	0.3	0.0	0.5
No response	1.7	1.6	1.6	1.9	1.5
0%	2.6	5.2	4.5	7.3	2.3
10%	4.5	8.4	7.7	8.9	3.8
20%	4.7	7.1	6.9	6.9	3.8
30%	5.1	6.6	7.2	8.8	4.5
40%	4.8	6.4	6.7	7.1	5.3
50%	7.6	9.2	8.0	8.4	6.4
60%	7.3	8.1	7.6	6.9	6.8
70%	9.1	8.8	8.2	8.5	8.7
80%	13.4	12.3	10.4	11.3	14.5
90% and up	38.9	26.1	30.9	23.7	42.1

quently (25%) supported entirely by parents. Near total aid is provided by parents in about 32% of cases at the men's colleges. It appears that biology and pre-med students rely more heavily on family aid than do students in the other disciplines (Table 6.4).

REPAYABLE LOANS

Although the response rate to the question of loan usage is poor for early graduates, the data suggest that reliance on loans increases between 1958 and 1967 (Table 6.5). The percentage of students receiving a small portion of aid in this form (10-30% of financial requirements) increases from 11% to nearly 23%. More substantial loan support, accounting for 40 to 100% of undergraduate expenses was obtained by 0.8% of the 1958 graduates and

Percentage of Support	1958 N = 1236 %	1959 N = 1292 %	1960 N = 1419 %	1961 N = 1589 %
Invalid	0.4	0.0	0.2	0.2
No response	19.5	17.3	14.6	15.6
0%	68.8	70.1	70.5	66.5
10%	6.3	7.5	9.0	10.9
20%	3.0	3.1	2.9	4.5
30%	1.2	1.0	1.6	1.0
40%	0.2	0.2	0.4	0.5
50%	0.5	0.4	0.6	0.4
60%	0.0	0.2	0.1	0.0
70%	0.0	0.1	0.1	0.0
80%	0.1	0.1	0.0	0.1
90% and up	0.0	0.2	0.0	0.3

2.8% of the 1967 graduates. The graduating class sizes increased by more than 30% for the sample considered in this study, whereas the number of students relying on loans for support has risen at a steeper rate than enrollment has increased. This differential is likely to be attributable to rising costs of college education, to the increasing ease of obtaining student loans for educational purposes, and to an increase in the enrollment of students from economic brackets that fall below the median income for this group of colleges. Further analyses, to assess the impact of each of these factors will be useful, particularly if based on rising parental incomes as an index of affluence.

Differences in the percentages of graduates in co-ed, male, and female colleges who support their education through loans are practically insignificant (Table 6.6). Of all science categories, pre-

TABLE 6.5

Percentage of Undergraduate Expenses Supported by Repayable Loans,
by Year of Graduation

1962 N = 1591 %	1963 N = 1724 %	1964 N = 1800 %	1965 N = 1919 %	1966 N = 1854 %	1967 N = 1942 %
0.2	0.4	0.2	0.2	0.2	0.4
14.4	15.3	12.3	12.2	10.3	9.9
68.2	65.7	66.3	62.9	63.2	63.7
10.5	10.3	10.5	12.5	13.0	12.1
3.6	4.7	5.7	6.6	7.5	7.4
1.8	2.3	2.7	3.1	3.5	3.9
0.6	0.9	1.1	1.3	1.4	1.2
0.4	0.2	0.6	0.6	0.5	0.8
0.1	0.1	0.2	0.3	0.2	0.4
0.1	0.1	0.2	0.3	0.2	0.1
0.1	0.0	0.1	0.1	0.1	0.2
0.1	0.2	0.2	0.1	0.1	0.1

TABLE 6.6

Percentage of Undergraduate Expenses Supported by Repayable Loans,
by College Type

Percentage of Support	Male Co-ed N = 7993 %	All-Male N = 3550 %	Female Co-ed N = 3152 %	All-Female N = 1700 %
Invalid	0.2	0.3	0.3	0.2
No response	13.6	13.1	13.9	15.5
0%	64.5	67.7	68.0	68.4
10%	10.8	9.4	10.4	11.8
20%	6.1	5.0	4.1	2.8
30%	2.6	2.7	2.1	0.8
40%	1.0	1.0	0.5	0.5
50%	0.6	0.6	0.3	0.1
60%	0.2	0.1	0.1	0.0
70%	0.1	0.1	0.1	0.0
80%	0.1	0.1	0.0	0.0
90% and up	0.1	0.1	0.1	0.0

TABLE 6.7

Percentage of Undergraduate Expenses Supported by Repayable Loans, by Major Subject

Percentage of Support	Biology N = 5996 %	Chemistry N = 3362 %	Math N = 3666 %	Physics N = 2045 %	Pre-med N = 1329 %
Invalid	0.3	0.3	0.2	0.2	0.2
No response	13.5	13.8	13.4	14.9	14.2
0%	68.0	64.8	64.4	62.1	73.7
10%	9.7	11.2	11.8	11.8	6.8
20%	4.6	6.0	5.7	6.5	1.8
30%	2.3	2.1	2.7	2.6	1.7
40%	0.8	0.7	0.9	1.1	0.8
50%	0.5	0.7	0.5	0.2	0.4
60%	0.1	0.2	0.2	0.2	0.1
70%	0.1	0.1	0.1	0.2	0.1
80%	0.1	0.0	0.1	0.1	0.1
90% and up	0.1	0.1	0.1	0.2	0.2

Percentage of Support	1958 N = 1236 %	1959 N = 1292 %	1960 N = 1419 %	1961 N = 1589 %
Invalid	0.2	0.1	0.2	0.3
No response	1.3	1.1	1.0	1.1
0%	42.1	45.4	45.5	43.0
10%	13.8	11.6	13.0	11.8
20%	9.1	7.8	9.0	8.6
30%	7.9	8.1	6.3	7.4
40%	6.2	5.0	5.0	5.9
50%	4.5	4.5	5.4	5.0
60%	1.9	2.6	1.8	2.8
70%	1.0	1.0	1.6	2.4
80%	0.4	1.3	1.3	1.5
90% and up	0.5	1.1	1.1	0.9

med students are least likely to use loans (Table 6.7). The percentage of individuals receiving loans in biology, math, physics, and chemistry, however, does not differ greatly across these groups (the range being between 18% and 23%).

SCHOLARSHIP AID

Tabulations on extent of scholarship aid are given in Tables 6.8, 6.9 and 6.10. Somewhat less than half (45%) of the students in each graduating class received some form of scholarship aid during their undergraduate years (Table 6.8). The extent of the aid, as a function of total financial requirements, is markedly consistent over the ten years considered. Regardless of the size of the class, a relatively stable portion of the graduates received some fixed ratio of support through scholarships. A slight oscillatory pattern is evident, with peaks occurring at 1961 and 1962 and again in

TABLE 6.8

Percentage of Undergraduate Expenses Supported by Scholarship Aid, by Year of Graduation

1962 N = 1591 %	1963 N = 1724 %	1964 N = 1800 %	1965 N = 1919 %	1966 N = 1854 %	1967 N = 1942 %
0.1	0.2	0.2	0.1	0.3	0.4
1.1	1.0	0.9	0.9	0.9	0.8
43.4	44.1	45.5	44.8	45.3	46.0
11.6	12.1	12.8	13.5	12.4	12.6
7.9	9.2	8.7	7.6	9.2	8.3
8.6	7.7	7.2	8.1	8.2	7.7
5.7	4.9	5.3	5.4	5.8	5.3
5.6	5.4	4.6	5.3	4.5	5.6
2.3	3.1	2.9	3.0	2.5	3.4
1.8	1.2	1.8	1.4	1.5	1.5
1.3	1.0	0.8	0.8	0.9	0.8
1.3	0.8	0.9	0.7	0.9	0.7

TABLE 6.9

Percentage of Undergraduate Expenses Supported by Scholarship Aid, by College Type

Percentage of Support	Male Co-ed N = 7993 %	All-Male N = 3550 %	Female Co-ed N = 3152 %	All-Female N = 1700 %
Invalid	0.2	0.2	0.2	0.3
No response	9.9	8.8	9.7	11.6
0%	42.9	45.4	45.9	48.6
10%	13.0	11.0	14.5	9.6
20%	9.4	8.1	8.5	5.4
30%	8.0	8.2	6.8	6.8
40%	5.5	6.0	4.8	5.3
50%	5.2	5.6	4.1	5.1
60%	2.5	2.9	2.4	3.8
70%	1.5	2.1	1.3	1.7
80%	0.9	1.2	1.0	1.2
90% and up	1.0	0.7	0.9	0.8

TABLE 6.10

Percentage of Undergraduate Expenses Supported by Scholarship Aid, by Major Subject

Percentage of Support	Biology N = 5996 %	Chemistry N = 3362 %	Math N = 3666 %	Physics N = 2045 %	Pre-med N = 1329 %
Invalid	0.2	0.2	0.2	0.3	0.2
No response	10.6	8.9	9.1	10.0	10.3
0%	51.3	38.5	39.9	36.7	55.1
10%	12.2	13.5	12.8	12.4	10.6
20%	7.6	9.3	9.3	10.1	5.9
30%	6.2	9.6	8.6	9.0	5.9
40%	3.8	7.0	6.4	6.7	4.3
50%	3.7	6.0	6.1	6.7	3.6
60%	2.1	2.8	3.3	3.8	1.7
70%	0.9	1.8	2.4	2.2	1.2
80%	0.8	1.4	1.0	1.3	0.7
90% and up	0.7	1.1	1.2	0.9	0.5

1967. The earlier fluctuations are probably due to post-Sputnik emphasis on the sciences.

When categories on percentage of aid are merged, an increase in dependency on scholarship aid is evident. Of 1967 graduates 12% received at least half of their financial support through scholarships. Only about 8% of the 1958 graduates acknowledged similar support. The rate of increases in substantial scholarship aid appears to be higher than the rate of increase in enrollment. The emphases of federal and private grant agencies have shifted during the past ten years, which may be responsible for this rate of funding.

Judging from Table 6.9, graduates of women's colleges are a bit less likely to have received scholarship aid than are any of the other sex-within-college-type categories. Men within co-ed colleges received aid slightly more frequently than did women. The percentage of aid, as a function of total requirements, is smaller for females than for males. Differences in nonresponse rate to this questionnaire item do not appear to be large enough to alter these inferences.

Pre-med and biology students are least likely to have received aid; 55 and 51% of respondents, respectively, indicated no scholarships (Table 6.10). The difference from students in other major fields is marked. Approximately half of the students in chemistry, physics, and math were recipients of scholarships.

WORK DURING SCHOOL YEAR

From 1958 to 1967, the percentage of students engaged in full- or part-time work during the academic year decreased (Table 6.11). About 45% of the earlier graduates acknowledged such employment, while only 36% of the 1967 graduates did so. Employment of students during the academic years 1958–1967 is moderated by the affluence of the students and the generally low wage level of students during college years.

The larger portion of students financing their education through

Percentage of Support	1958 N = 1236 %	1959 N = 1292 %	1960 N = 1419 %	1961 N = 1589 %
Invalid	0.3	0.3	0.3	0.4
No response	13.1	13.7	11.8	12.2
0%	41.7	42.7	45.5	46.8
10%	29.5	29.0	30.0	28.6
20%	8.6	8.5	7.9	7.1
30%	3.3	2.8	2.8	3.1
40%	1.9	1.1	0.7	0.8
50%	0.7	1.0	0.6	0.6
60%	0.2	0.5	0.1	0.1
70%	0.1	0.0	0.0	0.1
80%	0.1	0.2	0.1	0.1
90% and up	0.6	0.3	0.4	0.4

TABLE 6.12

Percentage of Undergraduate Expenses Supported
by Work During School Year,
by College Type

Percentage of Support	Male Co-ed N = 7993 %	All-Male N = 3550 %	Female Co-ed N = 3152 %	All-Female N = 1700 %
Invalid	0.3	0.2	0.2	0.4
No response	11.4	11.0	12.4	15.0
0%	46.6	49.0	49.7	56.5
10%	28.4	31.4	30.2	25.2
20%	8.5	5.7	4.7	2.3
30%	2.7	1.4	1.8	0.4
40%	0.8	0.8	0.4	0.2
50%	0.6	0.2	0.3	0.0
60%	0.2	0.1	0.0	0.0
70%	0.1	0.0	0.0	0.0
80%	0.1	0.0	0.0	0.1
90% and up	0.4	0.1	0.3	0.1

TABLE 6.11

Percentage of Undergraduate Expenses Supported
by Work During School Year,
by Year of Graduation

1962 N = 1591 %	1963 N = 1724 %	1964 N = 1800 %	1965 N = 1919 %	1966 N = 1854 %	1967 N = 1942 %
0.1	0.2	0.2	0.2	0.4	0.4
13.0	12.2	12.1	10.2	11.1	10.9
50.4	50.5	52.6	49.4	50.5	52.7
26.8	27.0	26.6	31.5	31.5	29.6
6.3	6.7	5.9	5.7	4.7	5.0
1.8	2.0	1.8	1.7	1.1	0.8
0.8	0.6	0.2	0.5	0.3	0.2
0.4	0.4	0.2	0.4	0.1	0.2
0.1	0.1	0.2	0.1	0.1	0.0
0.2	0.1	0.2	0.1	0.1	0.1
0.0	0.0	0.1	0.2	0.1	0.1
0.2	0.2	0.1	0.2	0.2	0.2

TABLE 6.13

Percentage of Undergraduate Expenses Supported
by Work During School Year,
by Major Subject

Percentage of Support	Biology N = 5996 %	Chemistry N = 3362 %	Math N = 3666 %	Physics N = 2045 %	Pre-med N = 1329 %
Invalid	0.3	0.2	0.2	0.4	0.2
No response	11.5	12.3	11.8	12.8	11.7
0%	48.9	47.4	49.6	44.7	55.8
10%	29.6	29.1	28.8	30.0	24.9
20%	6.4	6.6	6.4	7.2	5.1
30%	2.0	2.1	2.0	2.3	1.5
40%	0.6	1.0	0.4	1.0	0.2
50%	0.3	0.6	0.3	0.5	0.3
60%	0.1	0.2	0.1	0.2	0.1
70%	0.1	0.1	0.1	0.2	0.0
80%	0.1	0.2	0.1	0.1	0.0
90% and up	0.2	0.4	0.3	0.4	0.1

work were men in the co-ed colleges (Table 6.12). Of this group 42% had engaged in some form of work-study efforts; 11% had earned from 20 to 30% of their support by working during the school year. Men in all-male colleges were employed somewhat less frequently, and students in all-female colleges worked least during the academic year.

As shown in Table 6.13, pre-med students were least likely to have worked; more than 55% indicated that they had earned none of their educational expenses this way. Chemistry and physics majors appear to have provided for at least half of their expenses more frequently than any of the other categories. The number of students in all major field categories who earned 10% or less of their required expenses through work during the school year approached 80% in each group.

Percentage of Support	1958 N = 1236 %	1959 N = 1292 %	1960 N = 1419 %	1961 N = 1589 %
Invalid	0.2	0.0	0.0	0.1
No response	8.3	7.0	6.3	7.3
0%	19.4	19.4	19.5	21.4
10%	36.5	37.3	37.6	37.6
20%	17.6	19.4	21.1	19.1
30%	8.9	10.0	8.9	8.6
40%	5.1	3.3	3.5	2.9
50%	1.5	1.7	1.9	1.8
60%	1.2	0.9	0.8	0.4
70%	0.5	0.2	0.3	0.2
80%	0.5	0.2	0.1	0.1
90% and up	0.3	0.6	0.2	0.6

SUMMER WORK

The data relevant to summer work for educational expenses present a substantial contrast. Table 6.14 shows that almost 80% of all graduates had worked during their vacations in order to contribute to their own support. This proportion is remarkably stable for the ten years considered. Within years some variability in percentage of total support is evident. For example, the percentage of students earning at least half of their expenses in this way decreased from 4% to 2.2% over the years. Graduates who had earned from 10 to 20% of their expenses increased from 54% in 1958 to 62% in 1967. Again, the increasing fraction of the students contributing some small support and the concurrent decrease in larger self-support is probably a function of changes in low earn-

TABLE 6.14

Percentage of Undergraduate Expenses Supported by Summer Work, by Year of Graduation

1962 $N = 1591$ %	1963 $N = 1724$ %	1964 $N = 1800$ %	1965 $N = 1919$ %	1966 $N = 1854$ %	1967 $N = 1942$ %
0.0	0.1	0.2	0.1	0.2	0.3
7.5	6.1	7.0	6.7	6.3	5.8
22.4	21.6	22.3	23.7	21.8	21.0
36.8	39.0	40.3	40.2	41.2	41.2
20.0	19.0	17.2	17.5	18.9	20.3
7.8	9.7	8.6	7.8	7.2	7.5
2.6	2.8	2.2	2.7	2.8	2.0
1.4	1.0	1.2	0.6	1.2	1.3
0.6	0.2	0.4	0.3	0.1	0.4
0.4	0.4	0.1	0.3	0.2	0.1
0.1	0.0	0.1	0.0	0.0	0.1
0.4	0.2	0.2	0.1	0.2	0.3

TABLE 6.15

Percentage of Undergraduate Expenses Supported by Summer Work, by College Type

Percentage of Support	Male Co-ed N = 7993 %	All-Male N = 3550 %	Female Co-ed N = 3152 %	All-Female N = 1700 %
Invalid	0.1	0.2	0.2	0.2
No response	5.5	6.1	8.9	10.1
0%	16.5	19.0	28.9	35.4
10%	36.1	40.6	42.6	42.5
20%	21.8	21.7	14.4	8.6
30%	11.8	8.2	3.2	2.4
40%	4.3	2.5	0.9	0.5
50%	2.1	1.0	0.4	0.2
60%	0.8	0.3	0.1	0.0
70%	0.4	0.2	0.1	0.0
80%	0.2	0.1	0.0	0.0
90% and up	0.4	0.1	0.3	0.1

Percentage of Support	1958 N = 1236 %	1959 N = 1292 %	1960 N = 1419 %	1961 N = 1589 %
Invalid	0.2	0.1	0.1	0.1
No response	19.3	19.1	16.4	17.4
0%	64.6	65.9	66.6	65.9
10%	11.0	10.6	12.3	11.8
20%	3.0	2.6	3.2	2.6
30%	0.8	0.9	0.5	1.0
40%	0.4	0.1	0.1	0.5
50%	0.1	0.4	0.2	0.1
60%	0.1	0.2	0.1	0.2
70%	0.0	0.0	0.1	0.2
80%	0.1	0.1	0.0	0.0
90% and up	0.3	0.2	0.2	0.3

TABLE 6.16

Percentage of Undergraduate Expenses Supported by Summer Work,
by Major Subject

Percentage of Support	Biology N = 5996 %	Chemistry N = 3362 %	Math N = 3666 %	Physics N = 2045 %	Pre-med N = 1329 %
Invalid	0.2	0.1	0.1	0.1	0.2
No response	7.3	6.3	6.8	6.4	6.5
0%	23.1	19.2	22.5	18.0	21.4
10%	40.5	38.6	38.2	34.9	41.8
20%	17.2	21.1	18.8	21.7	17.9
30%	7.2	8.8	8.7	11.4	7.5
40%	2.4	3.2	2.6	4.6	2.5
50%	1.3	1.6	1.1	1.7	1.1
60%	0.4	0.7	0.4	0.6	0.5
70%	0.2	0.2	0.3	0.3	0.4
80%	0.1	0.1	0.1	0.1	0.1
90% and up	0.3	0.2	0.4	0.3	0.3

TABLE   6.17

Percentage of Undergraduate Expenses Supported by Personal Savings,
by Year of Graduation

1962 N = 1591 %	1963 N = 1724 %	1964 N = 1800 %	1965 N = 1919 %	1966 N = 1854 %	1967 N = 1942 %
0.0	0.2	0.1	0.2	0.2	0.3
16.4	16.9	14.9	14.2	13.9	13.5
67.8	67.3	66.4	67.9	66.8	67.3
11.9	11.7	14.1	13.9	14.4	15.2
2.3	2.1	2.5	2.0	3.0	2.0
0.8	1.2	1.0	1.2	0.5	0.6
0.3	0.3	0.3	0.2	0.3	0.4
0.2	0.2	0.3	0.1	0.3	0.4
0.1	0.1	0.1	0.1	0.1	0.2
0.0	0.1	0.0	0.2	0.2	0.1
0.1	0.1	0.2	0.1	0.1	0.0
0.2	0.0	0.2	0.1	0.3	0.2

TABLE 6.18

Percentage of Undergraduate Expenses Supported by Personal Savings,
by Major Subject

Percentage of Support	Biology N = 5996 %	Chemistry N = 3362 %	Math N = 3666 %	Physics N = 2045 %	Pre-med N = 1329 %
Invalid	0.2	0.1	0.1	0.1	0.2
No response	15.3	16.2	16.3	17.7	14.9
0%	66.9	67.1	66.8	65.2	67.6
10%	13.6	12.5	12.3	12.1	13.5
20%	2.3	2.4	2.7	3.0	2.2
30%	0.7	0.9	0.9	1.0	0.8
40%	0.3	0.2	0.3	0.4	0.2
50%	0.2	0.3	0.2	0.2	0.3
60%	0.2	0.1	0.1	0.2	0.1
70%	0.1	0.1	0.1	0.1	0.2
80%	0.1	0.1	0.1	0.1	0.0
90% and up	0.3	0.1	0.2	0.1	0.0

TABLE 6.19

Percentage of Undergraduate Expenses Supported by Personal Savings,
by College Type

Percentage of Support	Male Co-ed N = 7993 %	All-Male N = 3550 %	Female Co-ed N = 3152 %	All-Female N = 1700 %
Invalid	0.1	0.2	0.2	0.2
No response	15.9	14.9	16.4	17.8
0%	64.4	67.7	69.4	71.1
10%	14.2	13.6	10.9	8.6
20%	3.3	2.1	1.8	1.0
30%	1.1	0.7	0.5	0.2
40%	0.3	0.2	0.2	0.3
50%	0.2	0.2	0.1	0.4
60%	0.1	0.2	0.0	0.1
70%	0.1	0.1	0.1	0.0
80%	0.1	0.1	0.0	0.1
90% and up	0.2	0.1	0.3	0.3

ing power of the students and the more rapidly rising cost of a college education.

Graduates of women's colleges are least likely to have worked during the summer for educational expenses (Table 6.15). Men in the co-ed colleges are most likely to have done so (about 78% of the total group). Male-female differences are most pronounced, with men contributing to their educational expenses by working in summer jobs much more frequently than women.

No substantial differences among graduates within the major disciplines are evident (Table 6.16).

PERSONAL SAVINGS

Data on personal savings as a source of support further reinforce the impression of affluence in this student group. Approximately two-thirds acknowledge little or no self-support through savings. Less than 1% said savings helped to pay for more than half their educational expenses. The percentages are essentially the same regardless of the year or subject area (Tables 6.17 and 6.18). Fewer women than men acknowledged any self-support from personal savings. Men in co-ed colleges were most likely to have relied on personal savings (Table 6.19).

7

Transfers to and from Science Departments

The frequency of student transferral to, or attrition from, the science departments in college varies with institutional and curricular attributes, as well as student characteristics. A reasonable sketch of this aspect of student input at a departmental level is worthwhile for several reasons. First, relevant data can be utilized as a predictive base for manpower studies, perhaps to enhance the quality of forecasting the supply of physical science graduates. Second, data may be used more directly to estimate facility requirements. Finally, use of reliable data to derive appropriate policy decisions is essential to the regulation of attrition.

In this chapter we consider only the department chairman's perception of frequency and causes for attrition from the biology, chemistry, physics and mathematics departments in the liberal arts environment. For each department, estimates of actual percentage of acquisition and transferral (including attrition from college) are supplied. The percentages are based on the number of students initially enrolled in each major field of study. Chairmen were given six alternative causes, and requested to indicate whether these were of major importance, minor importance, or had no influence at all on the attrition rate from their department.

ATTRITION AND ACQUISITION OF STUDENTS

For each major field, estimates of departmental attrition rates are provided in Table 7.1. The general results are predictable—physics and mathematics department chairmen generally acknowl-

TABLE 7.1

Chairmen's Estimates on Percentage of Students
Who Left Science Departments

	Biology N = 41 %	Chemistry N = 41 %	Math N = 37 %	Physics N = 38 %
Less than 5%	46.3	34.2	18.9	15.8
5–9%	19.5	14.6	24.3	26.3
10–14%	9.8	24.4	10.8	10.5
15–19%	2.4	4.9	18.9	13.2
20–24%	4.9	2.4	8.1	7.9
25–29%	2.4	0.0	2.7	7.9
30–34%	4.9	7.3	0.0	5.3
35–39%	0.0	0.0	5.4	0.0
40–44%	0.0	4.9	2.7	2.6
45–49%	2.4	0.0	0.0	0.0
50% or more	0.0	2.4	2.7	2.6
Invalid	7.3	4.9	5.4	7.9

edge higher rates of attrition, and biology chairmen the lowest. Differences in attrition rates between the mathematics and physics departments are not significant. Chemistry chairmen place their groups at a level intermediate between the physics-math group and the biology departments. A half-dozen chairmen indicated a markedly high dropout frequency including more than 40% of initial student enrollment.

The rate at which students switch to a particular department varies as a function of major course requirements and departmental policies on time and inflexibility of major field declarations. Estimates of the resulting frequency in migration to science groups are given in Table 7.2. From the data it appears that biology and chemistry are more likely to acquire students from other departments, relative to physics and mathematics. Indeed, the acquisition rate for the former is somewhat greater than the

TABLE 7.2

	Biology $N = 41$ %	Chemistry $N = 41$ %	Math $N = 37$ %	Physics $N = 38$ %
	Chairmen's Estimates on Percentage of Students			
	Who Transferred to the Science Departments			
Less than 5%	39.0	31.7	46.0	50.0
5–9%	34.2	26.8	21.6	15.8
10–14%	22.0	22.0	13.5	10.5
15–19%	0.0	7.3	5.4	2.6
20–24%	0.0	2.4	0.0	5.3
25–29%	2.4	2.4	0.0	5.3
30–34%	0.0	0.0	0.0	0.0
35–39%	0.0	0.0	0.0	0.0
40–44%	0.0	2.4	2.7	0.0
45–49%	0.0	0.0	0.0	0.0
50% or more	0.0	0.0	0.0	0.0
Invalid	2.4	4.9	5.4	5.3

corresponding attrition rates presented in Table 7.1. Original student enrollments in most departments are augmented by from 5 to 15% because of transferrals from other departments.

There are a number of plausible alternative destinations for the students who drop out of the science curricula. These alternatives may include switching to another institution, switching to another major field of study, or discontinuing college entirely. Science department chairmen provided information on subsets of these possibilities. In order to permit assessment of science curriculum losses, they estimated the frequency of transferral to nonscience departments as well as to other scientific disciplines.

Examination of Tables 7.3 and 7.4 indicates that biology majors who transfer are somewhat more likely to choose a nonscience rather than a science program. In 73% of the biology departments less than 5% of the initial student enrollment switched to other science majors, while nearly 63% of the chairmen indicated a similar percentage of transfer to nonscience programs. In the physics department sample, approximately an equal proportion of chairmen (34%) acknowledged a negligible frequency of transfer (less than 5% of initial student enrollment) to either nonscience or science studies. Over 42% of the physics department respondents indicated that from 5 to 14% of the students who transferred chose other science departments, and nearly 48% of the chairmen said that approximately the same percentage of students switched to nonscience curricula. This slight differential in migration is not reflected for the chemistry departments, however; for these departments students apparently move to other science and nonscience curricula in equal proportions. Math students who transfer are more likely to change to a nonscience major.

TABLE 7.3

| | Chairmen's Estimates on Percentage of Students Who Transferred to Another Science Major | | | |
	Biology N = 41 %	Chemistry N = 41 %	Math N = 37 %	Physics N = 38 %
Less than 5%	73.2	43.9	56.8	34.2
5–9%	12.2	24.4	24.3	21.1
10–14%	7.3	12.2	10.8	21.1
15–19%	0.0	7.3	0.0	2.6
20–24%	0.0	4.9	0.0	7.9
25–29%	2.4	0.0	0.0	2.6
30–34%	0.0	2.4	0.0	2.6
35–39%	0.0	0.0	0.0	2.6
40–44%	0.0	0.0	0.0	0.0
45–49%	0.0	0.0	0.0	0.0
50% or more	0.0	0.0	0.0	0.0
Invalid	4.9	4.9	8.1	5.3

TABLE 7.4

	Chairmen's Estimates on Percentage of Students Who Transferred to a Nonscience Major			
	Biology N = 41 %	*Chemistry* N = 41 %	*Math* N = 37 %	*Physics* N = 38 %
Less than 5%	63.4	43.9	32.4	34.2
5–9%	2.4	24.4	29.7	34.2
10–14%	17.1	12.2	5.4	13.2
15–19%	4.9	4.9	5.4	2.6
20–24%	4.9	4.9	5.4	2.6
25–29%	2.4	2.4	0.0	2.6
30–34%	0.0	0.0	5.4	0.0
35–39%	0.0	0.0	2.7	0.0
40–44%	0.0	0.0	0.0	0.0
45–49%	0.0	0.0	2.7	0.0
50% or more	0.0	2.4	2.7	0.0
Invalid	4.9	4.9	8.1	10.5

There appear to be no real differences among departments in the extent to which students interrupt their liberal arts education and enroll at the universities (Table 7.5). Approximately 5 to 10% of students who do transfer from science departments of small liberal arts institutions enter the university systems, according to roughly 90% of the chairmen. The majority (80%) of the chairmen indicated, however, that any substantial transfer (more than 5%) generally occurs on a within-institution level rather than between the college and university.

In summary, the data suggest that departments are only somewhat more likely to acknowledge notable attrition rates than acquisition rates. For a small percentage of all departments, attrition rates are exceedingly high. For the majority, however, attrition is only in the 5 to 14% range. Notable rates of transfer to other science majors occur mostly (76%) in the physics department group. Chemistry, mathematics, and biology departments follow in that order. The transferrals to nonscience major fields are most likely in the math and biology departments. Biology has a relatively high retention.

REASONS FOR ATTRITION

The student's inability or unwillingness to continue in the specific course of study initially chosen can be interpreted as a function of a wide variety of possible reasons for his attrition. Department chairmen provided their views on the education-related causes of attrition (Tables 7.6 and 7.7). The chairmen were asked to rate as major, minor, or no influence the following proposed reasons for attrition: time commitment in laboratory work too demanding,

TABLE 7.5

	Chairmen's Estimates on Percentage of Students Who Transferred to a University			
	Biology N = 41 %	Chemistry N = 41 %	Math N = 37 %	Physics N = 38 %
Less than 5%	82.9	80.5	83.8	79.0
5–9%	4.9	9.8	5.4	5.3
10–14%	7.3	0.0	5.4	2.6
15–19%	2.4	0.0	0.0	0.0
20–24%	0.0	0.0	0.0	0.0
25–29%	0.0	0.0	0.0	0.0
30–34%	0.0	0.0	0.0	0.0
35–39%	0.0	0.0	0.0	0.0
40–44%	0.0	0.0	0.0	0.0
45–49%	0.0	0.0	0.0	0.0
50% or more	0.0	0.0	0.0	0.0
Invalid	2.4	9.8	5.4	13.2

TABLE 7.6

Chairmen's Appraisals of Factors in Student Attrition
as Major and Minor Influences,
by Major Subject

	Biology N = 41 %	Chemistry N = 41 %	Math N = 37 %	Physics N = 38 %
Time commitment in lab is too demanding				
Major influence	9.8	24.4	0.0	15.8
Minor influence	56.1	53.7	2.7	57.9
Mathematical competence for advanced courses lacking				
Major influence	7.3	48.8	81.1	65.8
Minor influence	39.0	39.0	13.5	31.6
Unable to comprehend advanced theories				
Major influence	7.3	31.7	89.2	36.8
Minor influence	58.5	46.3	5.4	52.6
Greater excitement and personal challenge				
Major influence	26.8	14.6	10.8	31.6
Minor influence	53.7	31.7	54.1	36.8

mathematical competence for advanced courses lacking, student's inability to grasp advanced theories, greater challenge elsewhere, disenchantment with space-age science, and attractiveness of departmental equipment or staff competence. Only the major reasons, as perceived by the chairmen, are presented.

The causes differ, in some cases substantially, depending on major department and on type of college. Laboratory time commitment is alleged to be a major deterrent most frequently by chemistry and physics departments, but is generally considered to be a minor influence, regardless of department or type of college. In the female college it is rated a major influence more frequently than in•the other college type categories.

Mathematical competence is generally held to be a major influence on a student's decision by all departments (especially the math department) except biology. This is not surprising since the importance of mathematics in the technical sciences is obvious, whereas math is not so crucial in the study of biology. By college type, mathematical competence is uniformly considered to be a major influence.

The student's general inability to comprehend advanced theories is not generally held to be a significant factor by biology,

TABLE 7.7

Chairmen's Appraisals of Factors in Student Attrition as Major and Minor Influences, by College Type

	All-Male N = 33 %	All-Female N = 27 %	Co-ed N = 98 %
Time commitment in lab is too demanding			
Major influence	6.1	22.2	14.3
Minor influence	48.5	40.7	41.8
Mathematical competence for advanced courses lacking			
Major influence	48.5	40.7	52.0
Minor influence	30.3	33.3	31.6
Unable to comprehend advanced theories			
Major influence	57.6	37.0	34.7
Minor influence	24.2	25.9	50.0
Greater excitement and personal challenge			
Major influence	15.2	33.3	19.4
Minor influence	27.3	44.4	49.0

chemistry, or physics chairmen, but is almost unanimously considered to be a major factor in the mathematics student's decision. This factor is also uniformly held to be major by chairmen in all college types. It seems to have more substantial significance in all-male colleges.

The greater excitement and personal challenge offered by another discipline is perceived to be a minor influence by all departments and college types by a majority of the chairmen queried.

Generally little or no importance is attached to the hypothesis that the "glamor of space-age science may have lost its luster," or that the department may lose students to another department because of more extensive or high-quality staff and equipment.

Although the alternatives to this question were limited, we can form some reasonable judgments about attrition rates, based on the chairmen's responses. The impact of various factors is differential, depending, in some cases, on the curriculum and college type. Math departments appear to lose students largely because of the student's lack of competence in mathematics or inability to understand more advanced theories. Physics and chemistry chairmen attribute loss of students to these causes less frequently, but do feel these to be major influences. The lack of expertise in mathematics courses seems to be most crucial in the co-ed and men's colleges. The most important consideration as far as biology students are concerned seems to be the greater excitement and personal challenge offered by other fields, according to the chairmen of this department.

The attractiveness of departmental equipment or staff competence, and the student's general disenchantment with space-age science, do not appear to be significant influences in determining student attrition rates.

8

Undergraduate Achievement

Within many college environments, the alternative methods for recognizing student achievement vary considerably. For example, civic or state achievement awards typically differ from grades and from Phi Beta Kappa awards insofar as more standardized criteria for selection of award candidates are applied in the latter. This heterogeneity of criteria makes comparison of individual achievements difficult, especially if sex or major studies categories are not made explicit. Meaningful inferences can be made, however, based on reasonably well-specified and uniformly applied types of awards. In this chapter, we consider several of these forms of recognition for science student performance: rank in graduating class, grade-point average within major studies coursework and overall average, traditional college graduation honors, and membership in honorary societies.

Rank and grades, of course, are frequently correlated. However, the rankings do provide some independent information insofar as they reflect a normative and ordinal standard with a wide range. Membership in honors groups frequently reflects other aspects of student behavior; for example, students' interests, interpersonal relations, and other social criteria applied in selection of candidates. The customary college graduation honors (*cum laude,* etc.) or within-department awards depend largely on grades and occasionally on individual research projects.

Data on each of these forms of recognition were provided by college registrars. Responses to the questionnaire survey (Appendix B2) are classified by type of institution, by year of graduation, and by major discipline.

NOTE: *The number of individuals reported in Tables 8.1 through*

	1958 N = 1304 %	1959 N = 1379 %	1960 N = 1484 %	1961 N = 1686 %
1–5% (Top)	8.0	8.5	8.4	7.2
6–10%	8.8	7.8	8.4	8.7
11–20%	12.9	12.8	11.7	10.6
21–30%	12.6	11.5	10.9	12.0
31–40%	11.7	11.2	11.0	10.4
41–50%	9.4	9.9	11.3	9.9
51–60%	8.3	8.9	8.8	9.3
61–70%	8.7	7.2	7.8	9.9
71–80%	6.4	8.3	8.3	8.8
81–90%	7.1	7.1	6.3	7.8
91–95%	3.7	3.1	3.4	2.5
96–100% (Bottom)	2.6	3.9	3.8	2.9

	1958 N = 1304 %	1959 N = 1379 %	1960 N = 1484 %	1961 N = 1686 %
4.00–3.75 (A or A+)	3.5	3.3	3.7	2.7
3.74–3.25 (A− or B+)	16.6	16.3	15.6	16.3
3.24–2.75 (B)	30.3	29.7	30.0	28.1
2.74–2.25 (B− or C+)	31.2	33.2	32.6	37.2
2.24–1.75 (C)	16.6	15.0	15.4	14.1
1.74–1.25 (C− or D+)	1.7	2.5	2.4	1.6
1.24–0 (D or less)	0.1	0.1	0.3	0.1

TABLE 8.1

		Graduates' Percentile Rank in Class, by Year of Graduation*			
1962 *N = 1620* %	*1963* *N = 1830* %	*1964* *N = 1899* %	*1965* *N = 2045* %	*1966* *N = 1865* %	*1967* *N = 1968* %
6.4	7.8	6.5	5.8	6.9	7.5
7.8	8.2	7.6	6.9	8.3	7.6
12.8	12.3	12.8	11.9	11.4	12.1
12.8	11.0	13.1	12.1	10.4	10.0
11.4	12.4	9.3	12.3	13.6	10.0
9.1	9.5	10.5	9.5	9.9	14.7
9.3	8.6	8.3	9.5	10.5	9.9
8.2	8.6	10.8	10.6	9.2	10.4
7.8	8.0	7.5	7.2	7.6	6.6
7.5	7.3	7.6	7.2	6.3	5.6
3.5	3.4	3.0	4.0	3.7	3.1
3.5	3.0	2.9	3.0	2.4	2.5

Note: Data provided by colleges' registrar offices, not by graduates.

TABLE 8.2

		Graduates' Overall Grade-Point Averages, by Year of Graduation*			
1962 *N = 1620* %	*1963* *N = 1830* %	*1964* *N = 1899* %	*1965* *N = 2045* %	*1966* *N = 1865* %	*1967* *N = 1968* %
2.4	2.7	2.6	2.0	2.3	2.2
16.3	15.4	16.8	15.9	15.9	16.1
31.7	32.6	32.1	31.9	32.4	35.1
34.9	34.7	34.3	35.9	36.4	34.0
13.0	13.2	12.6	13.2	12.0	11.4
1.6	1.2	1.4	0.9	0.9	1.1
0.1	0.2	0.3	0.2	0.2	0.1

Note: Data provided by colleges' registrar offices, not by graduates.

8.11 differs from the numbers reported in other chapters; the discrepancies result from differences in sources—that is, science graduates returning questionnaires versus (in this chapter) complete records provided by college registrar offices.

UNDERGRADUATE RANK AND GRADE-POINT AVERAGE

Table 8.1 includes data on the science graduates' rank in class. If one assumes comparable achievement criteria across all disciplines, about 40% of the science graduates would be expected to fall into the upper 40% of the total graduating classes. In fact, there is a slight positive bias (suggesting higher performance levels) in the science sample, with 50% of the group graduating within this level. From 6.4 to 8.5% of the science students graduated in the upper 5% of the classes with 6.9 to 8.8% of other graduates in the next highest category. There is a proportionate decrease in

	1958 N = 1304 %	1959 N = 1379 %	1960 N = 1484 %	1961 N = 1686 %
4.00–3.75 (A or A+)	8.4	7.8	7.9	8.0
3.74–3.25 (A— or B+)	19.9	20.4	21.5	20.2
3.24–2.75 (B)	30.1	27.9	26.8	28.0
2.74–2.25 (B— or C+)	24.9	27.9	28.0	28.1
2.24–1.75 (C)	14.0	13.1	13.0	13.8
1.74–1.25 (C— or D+)	2.6	2.9	2.6	1.7
1.24–0 (D or less)	0.0	0.1	0.3	0.2

science graduate representation within each of the groups included in the lower half of the class. These data are relatively stable across the ten-year period. The information on grade-point averages (Table 8.2) reflects much the same upward shift for the science graduates. There is a slight trend toward a smaller percentage of science graduates having grade-point averages of 3.74 and above, and a larger percentage above 2.74. Increases in the B— to C+ category amount to 3% of the graduate sample, while in the B category there is a 5% increase.

When only the grades for major coursework are considered in the computations (Table 8.3), the inferences are not altered substantially. That is, the percentage representation based on grades is quite similar to the data presented earlier. It appears that, on the average, students in this sample are more likely to perform well in their major coursework than they do in nonrelated coursework.

Table 8.4 contains pertinent rank data, classified by college

TABLE 8.3

Graduates' Major Coursework Grade-Point Averages, by Year of Graduation*					
1962 N = 1620 %	*1963* N = 1830 %	*1964* N = 1899 %	*1965* N = 2045 %	*1966* N = 1865 %	*1967* N = 1968 %
7.1	8.3	7.2	6.9	7.2	7.6
22.6	19.9	24.1	20.9	20.3	21.3
30.6	31.7	28.1	33.2	30.7	33.5
25.9	26.9	27.8	27.5	30.3	25.8
12.0	11.8	11.1	10.1	10.1	10.4
1.7	1.3	1.5	1.0	1.1	1.3
0.1	0.2	0.3	0.2	0.2	0.1

*Note: Data provided by colleges' registrar offices, not by graduates.

TABLE 8.4

	Male Co-ed N = 9152 %	All-Male N = 3396 %	Female Co-ed N = 3329 %	All-Female N = 1242 %
1–5% (Top)	6.2	8.0	8.0	10.1
6–10%	7.2	8.8	9.1	8.1
11–20%	10.9	13.0	14.7	11.6
21–30%	10.8	13.3	12.6	10.0
31–40%	10.9	11.5	13.3	8.9
41–50%	10.2	10.8	10.6	10.9
51–60%	9.3	9.0	8.8	9.9
61–70%	9.9	8.0	9.3	7.6
71–80%	8.4	6.9	6.2	7.9
81–90%	8.2	5.7	4.2	8.4
91–95%	4.0	2.8	2.0	3.5
96–100% (Bottom)	3.9	2.2	1.2	3.2

Graduates' Percentile Rank in Class, by College Type*

* *Note:* Data provided by colleges' registrar offices, not by graduates.

TABLE 8.5

	Male Co-ed N = 9152 %	All-Male N = 3396 %	Female Co-ed N = 3329 %	All-Female N = 1241 %
4.00–3.75 (A or A+)	2.5	2.5	4.0	0.7
3.74–3.25 (A− or B+)	15.5	14.6	20.1	14.3
3.24–2.75 (B)	30.6	24.0	39.7	38.0
2.74–2.25 (B− or C+)	34.9	36.7	30.3	38.4
2.24–1.75 (C)	15.4	18.0	5.7	8.1
1.74–1.25 (C− or D+)	1.2	3.6	0.3	0.6
1.24–0 (D or less)	0.0	0.7	0.0	0.0

Graduates' Overall Grade-Point Averages, by College Type*

* *Note:* Data provided by colleges' registrar offices, not by graduates.

type. For the graduates of coeducational colleges, the sex differential in ranks is most apparent, with about 31% of the women in the upper fifth of the total graduating class and less than 25% of the men ranking similarly. For the men's and the women's institutions, however, differences are negligible. Approximately the same percentage (30%) of male and female science graduates from single-sex schools rank in the upper fifth of their class.

Insofar as the data are based on unequal numbers of graduates within each category, and on overall average grades, inferences may be misleading. In the case of men's colleges, for example, the sample is large relative to the women's colleges graduates. There may be a smaller number of A grades for major coursework in the men's sample (Table 8.6), yet the graduating ranks may be relatively high because of a compensatory effect attributable to other course grades (Table 8.5). Grade-point averages within the major discipline are generally higher for female graduates, with females from co-ed institutions having higher grade-point averages than those from all-female colleges. Among men, it appears that graduates of co-ed colleges have higher grade-point averages than those from all-male colleges. The above comments are also consistent with the data on the major coursework grade-point averages presented by college type (Table 8.6).

Table 8.7 shows percentile rank in class by subject area and Table 8.8 contains grade-point average data, cross-tabulated against major field of study. Biology graduates fare poorly relative to the other groups' average performance in all coursework. From 6 to 9% fewer biology majors receive grade averages of B+ or better than do the students in the other fields. There are proportionally more of the biology graduates in the lower grade levels. One may have expected this group performance differential, perhaps based on observation of the S.A.T. scores of entering freshmen. However, the high overall grades of the pre-med students suggest that prediction of college performance from the S.A.T. scores must include recognition of the curricular requirements.

TABLE 8.6

Graduates' Major Coursework Grade-Point Averages,
by College Type*

	Male Co-ed N = 9152 %	All-Male N = 3396 %	Female Co-ed N = 3329 %	All-Female N = 1242 %
4.00–3.75 (A or A+)	7.6	5.8	9.3	8.0
3.74–3.25 (A— or B+)	20.0	19.1	24.5	26.8
3.24–2.75 (B)	30.8	24.1	34.6	31.5
2.74–2.25 (B— or C+)	27.8	29.4	24.7	25.5
2.24–1.75 (C)	12.4	17.1	6.4	7.3
1.74–1.25 (C— or D+)	1.4	3.9	0.5	1.0
1.24–0 (D or less)	0.1	0.7	0.0	0.0

* *Note:* Data provided by colleges' registrar offices, not by graduates.

TABLE 8.7

Graduates' Percentile Rank in Class,
by Major Subject*

	Biology N = 6387 %	Chemistry N = 3681 %	Math N = 3857 %	Physics N = 2246 %	Pre-med N = 952 %
1–5% (Top)	5.7	8.3	7.6	8.2	9.6
6–10%	4.9	8.8	10.6	10.8	7.5
11–20%	9.4	12.8	14.9	12.3	16.7
21–30%	10.8	12.3	11.2	12.0	14.4
31–40%	11.8	11.0	10.8	11.0	12.3
41–50%	11.3	10.1	9.4	9.4	12.9
51–60%	10.3	8.0	8.6	9.4	8.2
61–70%	11.5	8.9	7.6	7.4	6.1
71–80%	8.7	7.3	6.7	7.0	6.8
81–90%	8.2	6.2	6.5	7.0	4.0
91–95%	3.8	3.0	3.5	3.1	1.1
96–100% (Bottom)	3.6	3.3	2.6	2.5	0.5

* *Note:* Data provided by colleges' registrar offices, not by graduates.

TABLE 8.8

Graduates' Overall Grade-Point Averages,
by Major Subject*

	Biology N = 6387 %	Chemistry N = 3681 %	Math N = 3857 %	Physics N = 2246 %	Pre-med N = 952 %
4.00–3.75 (A or A+)	1.5	3.1	3.8	3.0	3.3
3.74–3.25 (A− or B+)	12.2	17.7	19.2	19.2	16.9
3.24–2.75 (B)	30.3	33.0	31.6	30.7	36.5
2.74–2.25 (B− or C+)	38.5	31.9	32.3	32.2	33.4
2.24–1.75 (C)	15.5	12.7	12.0	13.1	9.5
1.74–1.25 (C− or D+)	1.7	1.5	1.1	1.8	0.5
1.24–0 (D or less)	0.2	0.1	0.2	0.1	0.0

* *Note:* Data provided by colleges' registrar offices, not by graduates.

TABLE 8.9

Graduates' Major Coursework Grade-Point Averages,
by Major Subject*

	Biology N = 6387 %	Chemistry N = 3681 %	Math N = 3857 %	Physics N = 2246 %	Pre-med N = 952 %
4.00–3.75 (A or A+)	6.2	7.6	9.7	9.3	5.0
3.74–3.25 (A− or B+)	19.7	21.6	23.0	22.9	18.0
3.24–2.75 (B)	32.7	30.3	27.4	28.7	29.3
2.74–2.25 (B− or C+)	28.4	26.1	26.9	24.6	33.5
2.24–1.75 (C)	11.3	12.4	11.1	12.4	13.2
1.74–1.25 (C− or D+)	1.4	2.0	1.8	1.9	0.8
1.24–0 (D or less)	0.2	0.2	0.1	0.2	0.1

* *Note:* Data provided by colleges' registrar offices, not by graduates.

It may be true, for example, that the biology majors have fewer optional courses, or that they may be required to take more quantitative courses (which are presumably more difficult) than the pre-med students. Some evidence that this situation may result in differential predictability across groups is available from Table 8.9. This table contains data that suggest a pre-med–biologist–chemist grouping, and a physicist–mathematician grouping on performance in major coursework. That is, biology, pre-med and chemistry students perform at about the same average grade level, with physics and mathematics students on a different grade level.

OTHER FORMS OF RECOGNITION

Relatively poor records of subject-oriented honorary society membership are kept by the colleges in this sample. Lists of Phi Beta Kappa recipients were maintained most consistently, however. Table 8.10 contains percentages of total science student enrollment who received this award, classified by year of graduation and college type.

A most striking feature of these data is the high ratio of men's

| | 1958 | | 1959 | | 1960 | | 1961 | |
	N	%	N	%	N	%	N	%
Male Co-ed	701	9.0	744	9.1	810	8.2	896	9.6
All-Male	275	18.2	307	14.7	315	16.8	333	13.5
Female Co-ed	229	9.2	222	11.3	270	11.5	331	12.7
All-Female	98	9.2	105	6.7	88	6.8	126	6.4

college students who have been awarded the Phi Beta Kappa. Although this percentage appears to be declining somewhat over time, it is consistently higher than the data for the other college types. This statistic seems to contradict the data of Table 8.5, where it is indicated that graduates of all-male colleges have the smallest percentage of individuals with a B average or above. This contradiction reflects the latitude possible in applying selection criteria for such an award. Male graduates of co-ed institutions are slightly less likely to receive the award than female co-eds. There also appears to be a slight shift in the magnitude of percentages across time, with co-ed women becoming more likely to receive the award. In later years the ratio of recipients to science student enrollment is roughly the same for graduates of the women's colleges and female graduates of the co-ed colleges.

Data on college and departmental honors are usually more dependent on simple grade-point averages than are the Phi Beta Kappa awards. Table 8.11 reflects these annual forms of recognition within the institutional types. The percentage of graduates of women's colleges who receive honors is consistently higher than the percentage of recipients in any other type of college. Female

TABLE 8.10

Graduates Receiving Phi Beta Kappa, by Year of Graduation and College Type*

1962		1963		1964		1965		1966		1967	
N	%	N	%	N	%	N	%	N	%	N	%
875	8.3	962	10.2	1021	9.3	1102	8.4	966	8.8	1053	7.5
309	12.0	343	13.7	390	14.1	371	12.9	371	11.6	370	14.6
316	12.0	396	7.8	361	10.8	397	12.3	398	10.6	398	13.1
120	10.0	129	9.3	127	11.0	173	10.4	128	8.6	147	11.6

* *Note:* Data provided by colleges' registrar offices, not by graduates.

	1958 N = 701	1959 N = 744	1960 N = 810	1961 N = 896
Male Co-ed				
	%	%	%	%
Cum Laude	6.7	5.9	6.4	5.8
Magna Cum Laude	4.4	4.4	3.8	5.0
Summa Cum Laude	2.9	3.5	3.3	2.6
All-Male	N = 275	N = 307	N = 315	N = 333
	%	%	%	%
Cum Laude	13.5	15.3	14.9	11.4
Magna Cum Laude	4.4	5.9	5.1	4.5
Summa Cum Laude	2.2	1.0	3.8	1.2
Female Co-ed	N = 229	N = 222	N = 270	N = 331
	%	%	%	%
Cum Laude	10.5	11.3	7.4	10.9
Magna Cum Laude	3.5	3.6	5.9	5.7
Summa Cum Laude	0.9	2.3	4.1	3.3
All-Female	N = 98	N = 105	N = 88	N = 126
	%	%	%	%
Cum Laude	12.2	14.3	13.6	8.7
Magna Cum Laude	10.2	1.0	4.6	12.7
Summa Cum Laude	0.0	1.0	0.0	1.6

graduates of co-ed colleges receive a larger percentage of honors than the men graduating from these colleges. This substantiates the data presented earlier which implied that a larger percentage of female graduates of co-ed colleges had high grade-point averages than did the men. In early years students graduating from all-male colleges received only a slightly smaller percentage of honors than females from women's colleges. From 1961 on, however, the percentage of women receiving honors has exceeded the percentage of male recipients from 3 to 16%. There has been a slight increase in the overall percentage of females receiving honors from both the all-female colleges and the co-ed colleges.

TABLE 8.11

				Graduates Receiving Honors, by Year of Graduation and College Type*	
1962	1963	1964	1965	1966	1967
N = 875	N = 962	N = 1021	N = 1102	N = 966	N = 1053
%	%	%	%	%	%
8.1	5.6	6.8	8.6	7.4	7.0
3.5	5.0	4.6	3.5	4.2	3.4
2.5	2.8	2.7	2.0	3.1	2.9
N = 309	N = 343	N = 390	N = 371	N = 371	N = 370
%	%	%	%	%	%
12.9	13.7	15.4	14.0	12.4	14.9
4.9	2.9	4.4	4.3	4.6	5.7
2.6	1.2	3.3	1.1	1.6	1.4
N = 316	N = 396	N = 361	N = 397	N = 398	N = 398
%	%	%	%	%	%
12.7	7.1	6.7	6.1	8.5	8.3
4.1	3.8	4.4	6.3	5.3	5.5
2.9	2.8	5.3	2.5	2.3	4.0
N = 120	N = 129	N = 127	N = 173	N = 128	N = 147
%	%	%	%	%	%
23.3	16.3	20.5	19.1	20.3	17.7
12.5	10.9	11.0	8.7	9.4	6.8
0.0	3.1	0.8	1.2	1.6	4.1

* *Note:* Data provided by colleges' registrar offices, not by graduates.

9

Opinions of Science Department Chairmen

In this chapter and the following one, attention is restricted to professorial evaluation of the liberal arts environment, including the nature and extent of curricular requirements and science department orientations. Department chairmen and faculty in the liberal arts science programs constitute the basis for assessment. Insofar as responses are not biased systematically, the data reflect institutional conditions reasonably well.

One possible source of bias may be faculty "morale." Limited data on faculty attitudes have been obtained in order to permit examination of this issue. These data are presented in Chapter 10. Questionnaire samples for both faculty and department chairmen are exhibited in Appendix B.

CURRICULAR REQUIREMENTS

The distributed coursework required in liberal arts colleges may discourage or frustrate the student who is considering science as a major field of study. Department chairmen from chemistry, biology, mathematics and physics departments provided their opinions on this contention. As shown in Table 9.1, the vast majority of the members in each group reported no negative impact against a student's choice of the science department, attributable to distribution requirements. The largest number of affirmative responses occurred within the physics departments. Approximately 13% of the chairmen felt that general coursework requirements inhibited students in choosing physics as a major.

TABLE 9.1

Chairmen's Opinions on Liberal Arts Distribution Requirements as
Deterrents in Electing Majors in Science,
by Major Subject

	Biology N = 41 %	Chemistry N = 41 %	Math N = 37 %	Physics N = 38 %
Do you believe such requirements deter students who might otherwise elect major in department?				
Yes	2.4	7.3	2.7	13.2
No	95.1	90.2	97.3	81.6
Invalid	2.4	2.4	0.0	5.2
Would you favor reducing number of distribution requirements so majors in your department could be provided greater depth and breadth of preparation?				
Yes	17.1	14.6	13.5	13.2
No	80.5	80.5	83.8	81.6
Invalid	2.4	4.9	2.7	5.3

In addition to distribution of studies, there are, of course, other influences on a student's choice. Some of these were discussed earlier under the question of major influences in choice of the curriculum. Another influence in student decision may be the high ability requirements. In physics, for example, potential competition is reflected by the extremely high S.A.T. score averages for the enrolled students.

Curricular requirements may be altered, and some chairmen indicated that a reduction in distributed studies is desirable. Approximately 14% of all those questioned favored such a change. In this case, the biologists most frequently (17%) endorse changes that would presumably allow further concentration on major studies. No significant departure from these data is evident when the respondents are classified by college type (Table 9.2).

In summary, it appears that the large majority of department chairmen see no recruitment problem attributable to the demands of distributed studies. Only a small number endorse reduction of requirements in order to facilitate recruitment of students into science. This suggests a refusal by department chairmen to adjust traditional standards of education simply for the sake of recruiting more students.

TABLE 9.2

Chairmen's Opinions on Liberal Arts Distribution Requirements as Deterrents in Electing Majors in Science, by College Type

	All-Male N = 33 %	All-Female N = 27 %	Co-ed N = 98 %
Do you believe such requirements deter students who might otherwise elect major in department?			
Yes	12.1	3.7	5.1
No	84.9	96.7	91.8
Invalid	3.0	0.0	3.0
Would you favor reducing number of distribution requirements so majors in your department could be provided greater depth and breadth of preparation?			
Yes	24.2	18.5	10.2
No	75.8	70.4	86.7
Invalid	0.0	11.1	3.1

TABLE 9.3

Chairmen's Opinions on Principal Orientations of Departments

	First Choice N = 159 %	Second Choice N = 159 %
Graduate school preparation	63.5	26.4
Involve students in subject through research and/or project work	17.0	27.0
Provide adequate background for job entry in industry or business	3.8	20.1
Serve liberal arts program by providing most or all students with basic experience in discipline	12.0	13.8
Supplement majors in other areas with significant minor in math or science	0.0	6.9
Invalid	3.8	5.7

DEPARTMENTAL ORIENTATIONS

In order to assay the principal orientations of the mathematics and science departments, chairmen were requested to designate their first and second choice objectives from a given list of five alternative departmental aims or orientations: graduate school preparation, student "involvement" in the major discipline, business or industrial training, provision of basic science work for all liberal arts students, and provision of strong minor studies for other major study programs.

The majority (64%) of department chairmen indicated that graduate school preparation is their principal concern in determining administrative and educational emphasis (Table 9.3). Less than one-fifth of the respondents indicated that a first choice of departmental orientation was to involve the student in his major subject area through research or project work. Interestingly, none of the respondents indicated the primary objective to be supplementation of major studies in other disciplines with a strong minor in sciences or math. Less than 7% designated this as even a second choice, suggesting the rather low priority attached to minor studies.

Secondary choices for principal departmental orientations are mixed. The provision of adequate training for business or industrial employment was acknowledged by 20% of the chairmen. Student involvement in coursework and graduate school preparation were endorsed as secondary aims by 27% and 26% of department chairmen, respectively.

Summarizing, the department chairmen considered principal objectives to be training students for graduate school, and involvement of students in major subject matter through research. The efficacy of the former administrative aim is reflected in graduate school attendance data (postgraduate activities are reviewed in later chapters). Department chairmen place lower priorities on providing students for business or industrial employment, undergraduate major discipline experience, or strong minor studies in the science categories. Primary policy aims of department chairmen lead ultimately to further studies concentration by students. The science departments function more directly for production of graduate school candidates within the major.

GRADUATE SCHOOL ADMISSIONS CRITERIA

Department chairmen rendered opinions on the truth or fallacy of certain hypotheses relevant to graduate school admissions practices (Table 9.4). The purpose of the inquiry is to assess the extent to which chairmen perceive difficulties in sending students to graduate schools, and to examine possible reasons for the difficulties.

It is often alleged, for example, that the stringent grading practices in the liberal arts science curricula, relative to the

TABLE 9.4

Chairmen's Opinions on Effect of Liberal Arts Grading Standards on Graduate School Acceptance

	Biology $N = 41$ %	Chemistry $N = 41$ %	Math $N = 37$ %	Physics $N = 38$ %
Our grads have better chance of acceptance than those from less demanding schools despite oftentimes weaker academic average	51.2	36.6	18.9	13.2
Many of our grads have been penalized because grad schools do not recognize severe, discriminating character of our grades	7.3	2.4	0.0	7.9
Grad schools have tended to place greatest emphasis on G. R. E. and thus have paid little attention to our grading standards	2.4	0.0	0.0	10.5
Our recommendations carry sufficient weight to overcome any problem that might have arisen as result of academic average comparisons with those from other types of institutions	34.2	51.2	78.4	50.0
Invalid	4.9	9.8	2.7	18.4

practices at other types of institutions, result in penalization of many liberal arts graduates. In effect, the hypothesis that grades are lower for liberal arts science graduates because of the rigor in grading (resulting in reduced likelihood of admission to graduate school) is considered. The department chairmen in the current survey largely disagreed with this hypothesis. Across all departments, biology and physics chairmen endorsed the statement most frequently (7 and 8%, respectively). None of the mathematics chairmen, and only 2% of the chemistry chairmen felt this proposal was true.

When asked about positive biases in graduate admissions, the data shift across departments and generally reflect a more substantial number of affirmative responses. The hypothesis that liberal arts science graduates in this sample have a *better* chance of acceptance than those students from less demanding schools (with possibly higher grades) was supported by half of the biology chairmen. Physics chairmen were less optimistic (only 13% suggested that the situation is generally true). A substantial portion of chemistry (37%) and mathematics (19%) chairmen provided affirmative responses to this point.

Biologists and physicists most frequently voiced the opinion that graduate schools have emphasized Graduate Record Examination (G.R.E.) scores, rather than grades, in determining admission of students to advanced studies. The ratio is small (2.4 to 11%), but neither mathematics nor chemistry staff members felt that this was the case. It would be interesting to question graduate school admissions officers about their practices, insofar as the relative weight given to G.R.E. scores and grades may be important in determining curricular requirements or the contents of standardized tests.

To the extent that devices such as letters of recommendation are used in graduate school admissions criteria, the biases resulting from differentially rigorous undergraduate grading systems may be ameliorated. Department chairmen were requested to indicate their feeling about the following statement: "recommendations (from the department) carry sufficient weight to overcome any problem that might arise as a result of academic average comparisons between students from our institutions and those from other institutions." A substantial majority (78%) of the mathematics chairmen affirmed the statement. Both physicists and chemists were evenly divided as to the truth of this proposal. Chairmen from biology departments were most likely *not* to endorse this hypothesis; about two-thirds disagreed.

In conclusion, we find that a substantial portion of chairmen (particularly in the physics and mathematics departments) feel that rigorous grading practices at their institutions are not fully recognized in graduate school admission criteria. The impact of this is not ameliorated by attention to G.R.E. scores. However, recommendations have a definite impact insofar as they overcome problems arising from the differences in student evaluation. This appears to be particularly true for the mathematics departments while biology departments are largely unconvinced that this is the case.

10

Opinions of Science Department Faculty

Results of a special Science Faculty Questionnaire Survey (Appendix B4) permit a skeletal description of this group within the liberal arts environment. Inquiry is focused on biographic attributes and certain opinions relevant to the science faculty perceptions of their professional stature. Information on the latter was solicited primarily to allow some primitive assessment of faculty morale. The respondent group includes faculty members from chemistry (197 subjects), mathematics (188), biology (223) and physics (157) departments.

PERCEPTIONS OF PROFESSIONAL ESTEEM

Consider the esteem in which science faculty at the liberal arts institution are held by their counterparts at universities. Data were provided by the (liberal arts) science faculty in response to a question allowing four alternatives: there is substantial esteem afforded by membership in this group; personal esteem has no relation to the institution type; very slight negative trends exist; or a pronounced negative opinion is evident. Table 10.1 contains the results for this question, cross-tabulated against age categories. The following comments are made conditional on the sample of faculty who provided complete responses to the questionnaire. Note that the number of respondents in the most youthful category is suspiciously small, and the rate of nonresponse undermines the data's usefulness.

Overall, 15% said that science faculty at the liberal arts institution are held in substantial esteem by faculty members of universities. Generally, a much larger percentage of respondents suggested that the type of institution at which one is employed has nothing to do with the prestige afforded them. The majority of respondents under 49 years old acknowledged the feeling that slight, negative attitudes prevailed against faculty in the science departments of liberal arts colleges. The smallest overall proportion (12%) are of the opinion that university science faculty have a definite negative image of their colleagues in the liberal arts school. As age level increases, the percentage of those who felt that there is substantial esteem increases (from 7 to 38%) and there is a consequent decrease in the percentage who felt a negative attitude toward themselves.

Such impressionistic data can only be suggestive, but they are not encouraging. The majority of younger science faculty members appear to believe that other scientists' opinions are not complimentary. The changes in proportions across age categories may be a function of several variables. Younger faculty, with perceptions of poor professional image, may obtain jobs at other types of institutions. An alternative explanation of the change is that incumbent faculty gradually change their own perceptions of their image, and the alteration is in the morale-building direction.

On this latter point, some further information is available. Table 10.2 includes response frequencies tabulated against the number of years employed at the institution. It appears that faculty members with less service at the college perceive negative attitudes more frequently; only about 10% of those at the college less than 13 years feel that they are held in substantial esteem. On the other hand, approximately 30% of the respondents with greater tenure acknowledge that this is the case. As tenure increases, the proportion of those who believe that the type of institution is irrelevant to prestige also increases. There are some

	Under 25 N = 6 %	25–29 N = 124 %	30–34 N = 190 %	35–39 N = 125 %
Invalid	16.7	4.0	4.2	1.6
Liberal arts science faculty held in substantial esteem	50.0	7.3	9.0	6.4
Esteem has little or no relationship to type of institution in which one is employed	16.7	34.7	27.4	29.6
Noticeable though slight negative attitude exists within profession toward those in liberal arts science programs	16.7	41.1	45.8	46.4
Very definite negative image associated with teaching science in liberal arts colleges exists	0.0	12.9	13.7	16.0

puzzling peaks and depressions in percentages across tenure categories. These may be attributable to variations in average faculty member age across departments.

Across discipline categories, the physicist often perceives the least complimentary attitudes (Table 10.3). Approximately 19% felt a decidedly negative image of their own teaching positions by university science professors. In contrast, only 9% of the

TABLE 10.1

			Science Faculty's Perceptions of Esteem in Which Liberal Arts Science Faculty Is Held, by Age of Respondent		
40–44 *N = 94* %	*45–49* *N = 83* %	*50–54* *N = 52* %	*55–59* *N = 35* %	*60–64* *N = 35* %	*65 and over* *N = 26* %
2.1	2.4	0.0	8.6	2.9	7.7
17.0	19.3	25.0	20.0	37.1	38.5
30.9	26.5	38.5	45.7	34.3	34.6
38.3	36.1	26.9	14.3	22.9	15.4
11.7	15.7	9.6	11.4	2.9	3.9

mathematics respondents indicated this extreme case. Part of the difference in perceptions may be attributable to the prestige value of expensive or elaborate equipment in physics research, and the difficulty in acquiring or maintaining such equipment in the liberal arts college. It is also plausible to suspect that differences in opinion may be attributable to background history, such as the type of undergraduate institution attended by the

	Up to 3 years N = 281 %	4–6 N = 173 %	7–9 N = 76 %
Invalid	4.7	2.3	2.6
Liberal arts science faculty held in sub-stantial esteem	10.3	9.8	9.2
Esteem has little or no relationship to type of institution in which one is employed	30.6	29.5	36.8
Noticeable though slight negative attitude exists within profession toward those in liberal arts science programs	40.2	46.2	39.5
Very definite negative image associated with teaching science in liberal arts colleges exists	14.2	12.1	11.8

respondent. The highest proportion (nearly 16%) of faculty who felt they had a poor image occurs in the university graduate category (Table 10.4). Of the faculty who had attended liberal arts colleges, 10% felt this strongly; only about 7% of those from technical schools agreed to this response. From 35 to 40% of respondents in any category perceived slightly negative attitudes.

TABLE 10.2

			Science Faculty's Perceptions of Esteem in Which Liberal Arts Science Faculty Is Held, by Tenure of Respondent		
10–12 *N = 59* %	*13–15* *N = 49* %	*16–18* *N = 30* %	*19–20* *N = 22* %	*21–23* *N = 36* %	*24 and over* *N = 42* %
0.0	4.1	3.3	0.0	5.6	2.4
11.9	30.6	36.7	22.7	22.2	31.0
33.9	14.3	16.7	45.5	38.9	40.5
45.8	30.6	26.7	27.3	22.2	23.8
8.5	20.4	16.7	4.6	11.1	2.4

PERCEIVED DISCRIMINATION

Faculty members were asked to indicate whether or not they had encountered specific situations in which some persons or group had discriminated between liberal arts and university faculty, such that greater esteem or prestige was afforded the

TABLE 10.3

Science Faculty's Perceptions of Esteem in Which
Liberal Arts Science Faculty Is Held,
by Academic Area of Respondent

	Biology $N = 223$ %	Chemistry $N = 197$ %	Math $N = 188$ %	Physics $N = 157$ %
Invalid	2.2	3.1	4.3	3.8
Liberal arts science faculty held in substantial esteem	17.5	15.2	18.6	8.3
Esteem has little or no relationship to type of institution in which one is employed	33.2	29.4	29.3	31.2
Noticeable though slight negative attitude exists within profession toward those in liberal arts science programs	35.9	40.1	38.3	37.6
Very definite negative image associated with teaching science in liberal arts colleges exists	11.2	12.2	9.6	19.1

	Under 25 $N = 6$ %	25–29 $N = 124$ %	30–34 $N = 190$ %
Invalid	0.0	1.6	1.6
None	50.0	54.8	44.2
Once or twice	16.7	25.0	33.7
Several times	0.0	12.9	17.4
Many times	33.3	4.8	2.6
Any discrimination had favored those at liberal arts institutions	0.0	0.8	0.5

TABLE 10.4

Science Faculty's Perceptions of Esteem in Which
Liberal Arts Science Faculty Is Held,
by Type of Undergraduate Institution Attended by Respondent

	University N = 305 %	Technical Institution N = 42 %	Liberal Arts College N = 416 %
Invalid	3.9	4.8	2.9
Liberal arts science faculty held in substantial esteem	13.8	16.7	15.1
Esteem has little or no relationship to type of institution in which one is employed	31.5	35.7	29.8
Noticeable though slight negative attitude exists within profession toward those in liberal arts science programs	35.1	35.7	41.4
Very definite negative image associated with teaching science in liberal arts colleges exists	15.7	7.1	10.8

TABLE 10.5

Science Faculty's Perceptions of Discrimination
Against Liberal Arts Science Faculty,
by Age of Respondents

35-39 N = 125 %	40-44 N = 94 %	45-49 N = 83 %	50-54 N = 52 %	55-59 N = 35 %	60-64 N = 35 %	65 and over N = 26 %
0.8	0.0	2.4	1.9	2.9	0.0	0.0
41.6	52.1	44.6	53.9	65.7	71.4	76.9
34.4	30.9	30.1	23.1	14.3	25.7	11.5
19.2	13.8	19.3	13.5	14.3	2.9	0.0
3.2	1.1	2.4	7.7	2.9	0.0	3.9
0.8	2.1	1.2	0.0	0.0	0.0	7.7

latter. Four categories of response were provided on the questionnaire, the alternatives being focused on enumeration of such events: none, once or twice, several times, many times. A fifth alternative checklist item allowed the respondent to indicate that any discrimination had favored the liberal arts institution. Percentage data, classified according to age level of respondent, are given in Table 10.5.

Overall, the percentages of faculty who had encountered discrimination were: none, 51%; once or twice, 28%; several times, 15%; many times, 3.5%. Only 1.5% of the respondents indicated any instances of bias in favor of the liberal arts institution. The relevance of age to perception of biases against liberal arts science is confirmed in these data also. From 66 to 77 % of faculty members over 55 years of age acknowledged no cases of discrimination.

Across departmental categories, mathematics faculty appear to have been discriminated against in the fewest instances (Table 10.6); 58% responded that no biases were evident. Biology, physics, and chemistry response rates were not markedly different from

	Under 25 N = 6 %	25–29 N = 124 %	30–34 N = 190 %
Invalid	0.0	1.6	3.2
In general, universities offer stronger programs	50.0	29.0	32.1
Programs are about equal	50.0	51.6	51.1
In general, liberal arts colleges offer stronger programs	0.0	17.7	13.7

TABLE 10.6

Science Faculty's Perceptions of Discrimination
Against Liberal Arts Science Faculty,
by Academic Area of Respondent

	Biology N = 223 %	Chemistry N = 197 %	Math N = 188 %	Physics N = 157 %
Invalid	0.9	2.0	2.1	0.0
None	50.2	46.7	58.0	47.1
Once or twice	28.7	27.4	26.1	33.1
Several times	14.8	18.8	11.7	14.7
Many times	3.1	4.1	2.1	4.5
Any discrimination met favored those at liberal arts institutions	2.2	1.0	0.0	0.6

TABLE 10.7

Science Faculty's Perceptions of Strength of Liberal Arts
Undergraduate Science Programs,
by Age of Respondent

35–39 N = 125 %	40–44 N = 94 %	45–49 N = 83 %	50–54 N = 52 %	55–59 N = 35 %	60–64 N = 35 %	65 and over N = 26 %
2.4	4.3	2.4	3.9	11.5	5.7	7.8
28.0	22.3	30.1	25.0	28.6	25.7	30.8
48.8	46.8	50.6	50.0	40.0	51.4	46.2
20.8	26.6	16.9	21.2	20.0	17.1	15.4

one another. From 47 to 50% had experienced no situations in which discrimination against them was attributable to their institutional affiliation.

PERCEPTIONS OF THE SCIENCE PROGRAMS

Science faculty were asked to indicate whether they felt that the universities offer stronger undergraduate science programs than do the liberal arts colleges, whether such programs are about equal, or whether the liberal arts science programs are stronger. The data are presented here with respect to age and tenure levels of the respondents.

Averaging across all classifications given in Table 10.7, approximately 30% felt that universities offer stronger programs in undergraduate science programs. About 49% of the respondents considered the programs of liberal arts and university institutions to be about equal in quality, while 19% believe that the liberal arts curriculum is stronger than the universities.

Average responses are partly a function of age category;

	Up to 3 years N = 281 %	4–6 N = 173 %
Invalid	2.9	2.3
In general, universities offer stronger programs	32.4	34.1
Programs are about equal	50.5	45.1
In general, liberal arts colleges offer stronger programs	14.2	18.5

younger faculty endorsing the position that liberal arts colleges have the stronger programs. In terms of the faculty member's years of service at the college, it is those members with shorter tenure who affirm the same contention more frequently (Table 10.8).

More pronounced differences in nature of responses are evident in Table 10.9. Across academic disciplines, the mathematics faculty members most frequently render the opinion that university programs are superior. Similar opinion is expressed by fewer of the physics (28%), biology (24%) and chemistry (23%) respondents. The largest group affirming that study programs are stronger in liberal arts colleges were the biologists with 31% of these respondents taking this position. In contrast, only about 10% of the physics and math faculty agreed on the superiority of the liberal arts program while almost 20% of the chemistry members did so.

Some predictable differences are apparent when respondents are classified by type of undergraduate institution (Table 10.10). From 33 to 36% of university graduates and technical institute

TABLE 10.8

Science Faculty's Perceptions of Strength of Liberal Arts
Undergraduate Science Programs,
by Tenure of Respondent

7–9 N = 76 %	10–12 N = 59 %	13–15 N = 49 %	16–18 N = 30 %	19–20 N = 22 %	21–23 N = 36 %	24 and over N = 42 %
5.3	1.7	6.1	10.0	0.0	5.6	7.1
18.4	17.0	28.6	20.0	22.7	25.0	23.8
57.9	59.3	28.6	53.3	63.6	41.7	54.8
18.4	22.0	36.7	16.7	13.6	27.8	14.3

TABLE 10.9

Science Faculty's Perceptions of Strength of Liberal Arts
Undergraduate Science Programs,
by Academic Area of Respondent

	Biology N = 223 %	Chemistry N = 197 %	Math N = 188 %	Physics N = 157 %
Invalid	3.1	4.6	2.1	5.1
In general, universities offer stronger programs	24.2	22.8	40.4	28.0
Programs are about equal	41.7	54.8	47.3	55.4
In general, liberal arts colleges offer stronger programs	30.9	17.8	10.1	11.5

TABLE 10.10

Science Faculty's Perceptions of Strength of Liberal Arts
Undergraduate Science Programs,
by Type of Undergraduate Institution Attended by Respondent

	University N = 305 %	Technical Institution N = 42 %	Liberal Arts College N = 416 %
Invalid	4.0	9.5	2.4
In general, universities offer stronger programs	32.5	35.7	25.5
Programs are about equal	48.2	38.1	50.7
In general, liberal arts colleges offer stronger programs	15.4	16.7	21.4

graduates state that university programs are superior. About 26% of liberal arts graduates agree with this contention. The largest percentage of each group—48% of university graduates, 38% of technical graduates, and 51% of liberal arts graduates—felt that undergraduate programs were of equal quality.

SUMMARY

A majority of science faculty members included in this sample indicated that they perceive slightly negative attitudes held by university colleagues concerning the professional stature of the liberal arts faculty. Substantial portions of the faculty consider the image of science faculty at the liberal arts college to be rather poor. The feelings are most frequently expressed by younger faculty members and/or teachers with fewer years experience at the institution. Although the majority of respondents stated that they had experienced no discrimination as a result of their liberal arts affiliation, nearly one-third alleged they had indeed encountered instances of bias. Younger faculty and faculty with shorter tenure at their college were somewhat more likely to affirm superiority of the liberal arts science programs over the science programs at universities. Nearly half of all respondents perceived no real differences in quality between university and liberal arts studies.

It appears that a substantial proportion (although a minority) of liberal arts science faculty perceive themselves as having an unfavorable stature as afforded them by their university colleagues. Younger faculty members are a bit more likely to feel that this is the case. What impact these attitudes may have on behavior is, of course, an important issue. Higher attrition rates may be a function of the dissatisfaction implied by these data. Or, difficulty in recruitment of able teachers may be a result of the existence of such attitudes. A within-institution survey coupled with records of turnover rate may be an effective way to evaluate such hypotheses.

An additional important question is whether such opinions have an impact on students in the liberal arts colleges. To the extent that faculty members believe liberal arts programs to be at least as strong as university study, undergraduates may be encouraged in their own efforts and perhaps also in considering graduate studies. While younger faculty perception of stature may be low, their regard for the liberal arts program may be relatively high and may also be apparent to students. Conjectures such as these are impossible to verify except through in-depth study of the local campus environment, or more detailed analysis of the current data.

11

Graduate School Attendance
and Achievement

Within each graduating class, what proportion of science students have chosen to attend graduate school and have received a graduate degree? What are their areas of specialization? How well are the students performing in their graduate studies?

Responses to these inquiries were provided by science graduates who are currently enrolled, or who have been enrolled, in graduate studies. Tables 11.1 and 11.2 indicate the proportions of graduates (for each graduating class from 1958 to 1967) in each of these subdivisions, categorized by college type. More than half the graduates during this period have acquired at least one graduate degree (Table 11.1). The trend of current enrollments from 1958 to 1967 suggests that a substantial percentage of these students will be completing degree requirements. It appears that the graduate school orientation of department chairmen (described earlier) is being reflected markedly in the postgraduate activities of former students.

More specific information on the nature of graduate degrees being sought is given in Table 11.2. Master's degrees and doctorates, as well as professional awards, are considered in the context of college type and year of graduation. Note that there may be considerable overlap between the categories of master's and doctorate recipients. Assuming that fewer than 10% of the students are allowed to complete Ph.D. requirements without obtaining a master's, about half the students in the master's category will have stopped their graduate education at this level.

The most evident sex differentials occur in the categories of medically related degrees, the Ph.D., and law degress. The largest proportion of Ph.D.'s are men from co-ed institutions.

	1958	1959	1960
Male Co-ed	$N = 595$	$N = 636$	$N = 685$
Master's (M.A., M.S.)	32.8%	30.8%	33.4%
Doctor's (Ph.D., Ed.D.)	21.5	21.4	22.0
Medical (M.D., D.D.S., D.V.M.)	24.7	24.7	25.0
Law (L.L.B., J.D.)	1.2	1.4	0.7
Divinity (B.D.)	0.0	0.0	0.0
Nursing (R.N.)	0.2	0.8	0.7
Other	1.7	0.8	1.9
All-Male	$N = 275$	$N = 297$	$N = 329$
Master's (M.A., M.S.)	28.7%	26.3%	26.8%
Doctor's (Ph.D., Ed.D.)	17.5	15.2	19.5
Medical (M.D., D.D.S., D.V.M.)	45.1	47.1	43.8
Law (L.L.B., J.D.)	1.1	1.4	2.1
Divinity (B.D.)	0.7	0.7	0.0
Nursing (R.N.)	0.0	0.0	0.0
Other	2.6	2.0	0.0
Female Co-ed	$N = 227$	$N = 211$	$N = 253$
Master's (M.A., M.S.)	22.0%	27.0%	24.5%
Doctor's (Ph.D., Ed.D.)	5.7	6.2	6.3
Medical (M.D., D.D.S., D.V.M.)	4.4	2.8	3.2
Law (L.L.B., J.D.)	0.0	0.0	0.4
Divinity (B.D.)	0.4	0.0	0.0
Nursing (R.N.)	0.4	0.5	0.8
Other	3.1	3.8	1.6
All-Female	$N = 136$	$N = 146$	$N = 152$
Master's (M.A., M.S.)	26.1%	24.0%	25.0%
Doctor's (Ph.D., Ed.D.)	10.9	8.9	8.6
Medical (M.D., D.D.S., D.V.M.)	6.5	7.5	3.3
Law (L.L.B., J.D.)	0.0	0.7	0.0
Divinity (B.D.)	0.0	0.0	0.0
Nursing (R.N.)	0.7	0.0	1.3
Other	2.9	2.7	2.6

TABLE 11.1

		Advanced Degrees Earned by Graduates, by College Type and by Year of Graduation				
1961	*1962*	*1963*	*1964*	*1965*	*1966*	*1967*
$N = 775$	$N = 790$	$N = 833$	$N = 871$	$N = 947$	$N = 893$	$N = 953$
31.7%	33.2%	32.3%	28.8%	23.1%	16.7%	4.0%
19.6	17.2	12.1	4.8	0.3	0.1	0.0
22.5	22.0	24.6	20.0	1.1	0.6	0.5
2.2	0.8	1.1	0.7	1.8	0.0	0.0
0.9	0.1	0.0	0.0	0.0	0.0	0.0
0.1	0.0	0.6	0.7	0.1	0.0	0.0
1.8	1.0	1.0	0.8	1.1	0.5	0.2
$N = 346$	$N = 330$	$N = 348$	$N = 414$	$N = 391$	$N = 399$	$N = 410$
24.6%	27.3%	21.3%	27.3%	17.4%	10.8%	1.2%
19.1	13.9	12.9	2.7	0.5	0.0	0.0
35.6	37.0	38.8	28.7	0.3	0.8	0.0
1.2	0.6	1.2	0.7	2.6	0.3	0.0
0.3	0.3	0.6	0.2	0.0	0.3	0.0
0.0	0.0	0.0	0.0	0.0	0.0	0.0
1.2	0.9	0.6	0.2	1.0	0.3	0.2
$N = 305$	$N = 307$	$N = 358$	$N = 347$	$N = 367$	$N = 375$	$N = 394$
27.2%	28.7%	24.6%	20.8%	21.0%	10.9%	1.8%
5.9	3.9	3.4	0.9	0.0	0.0	0.3
6.6	5.2	4.2	2.3	0.8	0.0	0.3
0.0	0.0	0.3	0.3	0.0	0.0	0.0
0.0	0.0	0.0	0.0	0.3	0.0	0.0
0.7	0.3	0.0	1.4	0.3	0.3	0.0
3.9	4.6	3.4	2.6	4.9	2.4	1.0
$N = 163$	$N = 164$	$N = 185$	$N = 168$	$N = 212$	$N = 187$	$N = 185$
23.3%	32.9%	34.6%	25.0%	16.5%	20.3%	2.2%
9.8	3.1	3.8	2.4	0.0	0.0	0.0
10.4	10.4	6.0	7.7	0.0	0.5	0.0
0.6	0.0	0.0	1.2	0.0	0.0	0.0
0.0	0.0	0.0	0.0	0.0	0.0	0.0
0.6	0.0	0.5	0.0	0.0	0.5	0.0
2.5	0.6	2.2	0.6	1.4	1.1	0.0

	1958	1959	1960
Male Co-ed	$N = 595$	$N = 636$	$N = 685$
Master's (M.A., M.S.)	4.2%	5.2%	7.3%
Doctor's (Ph.D., Ed.D)	6.6	6.9	8.5
Medical (M.D., D.D.S., D.V.M.)	0.2	0.3	0.4
Law (L.L.B., J.D.)	0.3	0.3	0.0
Divinity (B.D.)	0.0	0.0	0.0
Nursing (R.N.)	0.2	0.0	0.0
Other	0.2	1.1	0.4
All-Male	$N = 275$	$N = 297$	$N = 329$
Master's (M.A., M.S.)	4.4%	5.4%	4.0%
Doctor's (Ph.D., Ed.D.)	4.4	5.7	7.3
Medical (M.D., D.D.S., D.V.M.)	0.4	0.0	0.6
Law (L.L.B., J.D.)	0.4	0.7	0.0
Divinity (B.D.)	0.0	0.0	0.0
Nursing (R.N.)	0.0	0.0	0.0
Other	1.5	0.3	0.3
Female Co-ed	$N = 227$	$N = 211$	$N = 253$
Master's (M.A., M.S.)	4.0%	5.2%	5.9%
Doctor's (Ph.D., Ed.D.)	2.2	3.8	4.0
Medical (M.D., D.D.S., D.V.M.)	0.4	0.0	0.0
Law (L.L.B., J.D.)	0.4	0.0	0.4
Divinity (B.D.)	0.0	0.0	0.0
Nursing (R.N.)	0.0	0.0	0.0
Other	1.3	0.5	0.4
All-Female	$N = 136$	$N = 146$	$N = 152$
Master's (M.A., M.S.)	5.1%	4.8%	2.6%
Doctor's (Ph.D., Ed.D.)	3.6	5.5	5.9
Medical (M.D., D.D.S., D.V.M.)	1.5	0.0	0.7
Law (L.L.B., J.D.)	0.0	0.0	0.0
Divinity (B.D.)	0.0	0.0	0.0
Nursing (R.N.)	0.0	0.0	0.0
Other	0.0	0.0	0.0

TABLE 11.2

Advanced Studies Under Way by Graduates,
by College Type and by Year of Graduation

1961	1962	1963	1964	1965	1966	1967
N = 775	N = 790	N = 833	N = 871	N = 947	N = 893	N = 953
7.4%	9.1%	8.6%	10.5%	9.9%	14.9%	22.9%
9.8	14.1	20.3	28.8	30.1	27.4	25.5
0.3	1.5	2.4	4.7	23.8	22.8	26.3
0.7	0.5	0.5	0.6	1.0	1.1	1.2
0.0	0.0	0.0	0.0	0.0	0.0	0.4
0.0	0.1	0.0	0.1	0.3	0.7	0.1
0.9	0.3	0.2	1.2	0.4	0.7	0.7
N = 346	N = 330	N = 348	N = 414	N = 391	N = 399	N = 410
4.6%	4.9%	6.3%	4.4%	10.0%	11.3%	14.4%
7.2	14.6	12.4	27.1	22.5	20.8	19.3
0.9	0.3	1.4	8.7	38.9	45.6	48.5
0.6	1.2	1.2	1.0	1.3	1.3	1.7
0.0	0.0	0.0	0.2	0.8	0.0	0.5
0.0	0.0	0.0	0.0	0.0	0.0	0.0
0.3	0.6	0.3	0.0	0.5	0.5	0.5
N = 305	N = 307	N = 358	N = 347	N = 367	N = 375	N = 394
5.3%	7.5%	11.5%	11.5%	10.9%	16.5%	25.1%
4.9	7.2	9.2	7.2	14.2	14.1	10.9
0.3	0.3	0.6	0.9	4.6	4.8	6.6
0.0	0.0	0.0	0.0	0.0	0.3	0.0
0.0	0.3	0.0	0.0	0.0	0.0	0.0
0.0	0.0	0.0	0.0	0.0	0.5	1.0
0.3	1.0	0.0	0.6	0.5	1.1	2.5
N = 163	N = 164	N = 185	N = 168	N = 212	N = 187	N = 185
6.1%	6.7%	7.6%	7.1%	14.6%	8.6%	18.9%
6.1	11.0	15.1	15.5	12.7	19.8	17.3
1.2	1.8	0.5	4.2	6.6	5.4	9.7
0.0	0.0	0.0	0.0	1.4	1.1	1.1
0.0	0.0	0.0	0.0	0.0	0.0	0.0
0.0	0.0	0.5	0.0	0.0	0.0	0.0
2.5	0.6	1.6	0.0	0.5	3.7	0.5

Graduates of the men's colleges are somewhat less likely to acquire a doctorate. Women's college graduates are next in this ordering with their representation in the Ph.D. category being about half that of men's college graduates. The co-ed colleges produce the smallest fraction of females who acquire a Ph.D.

Co-ed colleges also rank high in production of students who eventually enter the medical professions, but male college productivity in this area is substantially higher. Note that formal pre-med programs are not the only source of candidates for medically related degrees. Substantial numbers come from other major programs in the liberal arts studies. Some further analysis is required to pinpoint these other sources accurately. From 6 to 10% of the graduates of women's colleges become M.D.'s. Fewer (4 to 6% through the ten-year period) women from co-ed institutions achieve this level. A small but consistent fraction of women graduates of the co-ed institutions receive R.N.'s. The corresponding ratio for the women's colleges is much smaller and more sporadic.

The legal profession attracts only a small proportion of science graduates. The percentage of men choosing a law degree is fairly consistent over the ten-year period considered, but the attainment of law degrees by women deviates sharply from year to year, the lowest percentages occurring in this category particularly among women from the co-ed colleges.

Divinity degrees are awarded infrequently. The majority of recipients are from the men's colleges. A very small percentage of men and women enter divinity schools after graduation from science programs at the co-ed institutions.

SPECIALIZATION IN GRADUATE STUDIES

Consider now the respondents' programs of graduate study. The data in Table 11.3 are interesting insofar as they show a consistency with which students pursue specific fields of study. These data also show the magnitude and directions of shift in the graduates' interests. One can infer from these data that frequency regarding changes in major discipline by students entering graduate school is a function of the major discipline pursued in undergraduate school. Relative to the science classifications, the students who are *most* likely to choose advanced studies in different programs after graduation are those who had been enrolled in undergraduate biology and mathematics curricula. The other specialties to which mathematics students finally migrated include (roughly in order of frequency): education, business, physics, law and the social sciences. Biology students are most likely to change to medicine and other health professions.

Pre-med students are most consistent in their choice of postgraduate training. Nearly three-quarters of them choose medicine, and an additional 17% are divided between other health-related fields and biology. The pre-med students choose education (1.7%) and law (0.9%) as alternative programs.

Physics and chemistry majors are an intermediate grouping with respect to alteration of study emphasis. Approximately half of the graduates in each of these disciplines further pursue their undergraduate course of study. Of the physicists who choose other graduate fields, most entered engineering programs (13%). Business or commerce graduate schools acquired 6.4% of the remaining physics graduates, while small but notable ratios of these students entered fields closely allied with physics: mathematics, interdisciplinary science, or other physical sciences. Chemistry graduates most frequently choose medicine as an alternative graduate emphasis. This situation probably reflects the undergraduate options available to the student enrolled at an institution with no pre-med curricula per se.

The flexibility of background requirements implied by medical school selection criteria is also influential. Graduate studies of interdisciplinary science, business and education are chosen much

	Biology N = 4463 %
Invalid	1.8
Biology	30.4
Chemistry	0.9
Engineering	0.0
Health professions	7.6
Medicine	41.4
Mathematics	0.3
Physics	0.1
Interdisciplinary science	1.7
Other physical sciences	0.4
Agriculture	0.4
Architecture	0.2
Business or commerce	1.5
Education	6.9
Law	1.5
Mathematical subjects	0.0
Philosophy	0.0
Psychology	1.2
Religion	0.6
Social sciences	1.1

less frequently than medicine but the data imply a significant number of changes in aspirations.

ACADEMIC PERFORMANCE

Graduate school grade-point averages are a convenient, though primitive, index of the extent to which graduates have committed themselves to advanced studies. The retrospective reports of

TABLE 11.3

| | Specialization in Graduate School, by Major Subject* | | |
Chemistry $N = 2755$ %	Math $N = 2271$ %	Physics $N = 1632$ %	Pre-med $N = 1210$ %
1.2	1.6	1.0	1.1
2.4	0.5	1.5	8.9
49.9	0.4	0.8	1.3
1.2	2.9	13.3	0.1
2.2	0.3	0.3	8.1
24.8	2.6	2.0	74.2
0.4	42.9	3.3	0.3
0.4	3.6	53.8	0.2
4.5	1.0	3.8	0.3
0.9	1.1	3.6	0.2
0.0	0.0	0.0	0.0
0.0	0.7	0.3	0.1
4.0	9.5	6.4	0.8
3.4	10.2	2.1	1.7
1.2	3.0	1.3	0.9
0.3	11.5	1.9	0.1
0.4	1.0	0.8	0.0
0.5	1.5	0.7	0.6
0.4	0.7	0.7	0.2
0.7	3.0	1.4	0.2

*Note: These data include those who have received postgraduate degrees as well as those who have merely attended graduate school.

course grades in the current study were furnished by graduates who have been enrolled or who are currently enrolled in graduate study programs. The statistical data summarizing these reports are furnished in Tables 11.4 and 11.5. Assuming that grading practices do not differ substantially across graduate schools and departments, the pooled grade-point average can function as a rough index of student achievement level.

In Table 11.4, the proportions of students within each achieve-

	1958 N = 904 %	1959 N = 953 %	1960 N = 1065 %	1961 N = 1201 %
Invalid	0.6	0.5	0.9	0.5
A or A+	7.7	9.0	8.4	10.5
A−	20.8	20.0	22.0	20.6
B+	26.3	27.3	25.7	26.9
B	25.3	23.2	24.1	22.3
B−	10.3	11.1	9.5	10.8
C+	6.0	5.9	5.4	5.7
C	2.4	2.2	3.1	2.2
C−	0.4	0.6	0.6	0.4
D	0.1	0.1	0.3	0.0

ment level are furnished for each of the years 1958 through 1967. The fractions within each grade level are moderately stable over time and show only a slight trend of either increase or decrease depending upon the level. The range of percentages within each grade level are small. Stability of grading practices and relative quality of students over time is suggested by the narrow limits of variation: 8 to 11% in the A–A+ category; 20 to 26% (A−); 22 to 27% (B+); 20 to 25% (B); 9 to 12% (B−); 9 to 10% (C's). A negligible ratio of students performed at the D level. The largest single level of achievement includes the B+, B, B− categories, with over half the graduates in each year receiving these grades.

The minor instability in grade-point average over the years is likely to be a function of heterogeneity in both graduate school grading practices and enrollment.

TABLE 11.4

			Grade-Point Averages in Graduate School, by Year of Graduation*		
1962 *N = 1215* %	*1963* *N = 1326* %	*1964* *N = 1363* %	*1965* *N = 1390* %	*1966* *N = 1287* %	*1967* *N = 1269* %
0.5	0.2	0.3	0.1	0.0	0.2
9.8	11.1	10.1	10.5	9.6	10.1
22.3	24.0	26.4	21.9	22.1	18.4
26.1	26.4	25.9	25.7	24.0	22.4
22.6	20.8	20.7	21.1	24.0	25.4
10.1	9.7	8.7	11.3	10.0	12.2
6.0	4.4	5.0	5.4	5.9	6.1
1.7	2.6	2.2	2.9	3.0	4.1
0.6	0.4	0.5	0.8	1.0	0.9
0.2	0.2	0.1	0.3	0.2	0.2

* *Note:* These data include those who have received postgraduate degrees as well as those who have merely attended graduate school.

Another interesting view of the data clarifies sex differences in achievement (Table 11.5). Using undergraduate grades as a basis for judgment, one might expect women to perform at somewhat higher levels than the male graduates (see Chapter 8). For female graduates of co-ed institutions the expectation is justified. The distribution of grades for this group is skewed in the A+ direction, relative to men or women graduates of other types of institutions. The differences in grades among male and female graduates of co-ed institutions do not appear to have any practical significance.

The graduate grades of female co-ed graduates are consistent with undergraduate achievement insofar as performance of this group is slightly higher than other groups. The relatively low grade-point averages of graduates of men's institutions are also consistent with data on undergraduate performance.

TABLE 11.5

	Grade-Point Averages in Graduate School, by College Type*			
	Male Co-ed N = 6426 %	All-Male N = 2980 %	Female Co-ed N = 1646 %	All-Female N = 915 %
A or A+	9.3	8.5	14.5	9.3
A—	21.8	19.4	27.8	26.7
B+	25.8	22.7	28.1	28.5
B	22.4	26.1	18.3	22.6
B—	10.7	13.0	6.4	6.6
C+, C, C—, D	9.4	12.3	4.3	6.6

*Note: These data include those who have received postgraduate degrees as well as those who have merely attended graduate school.

The overall graduate school averages are more frequently higher than major coursework averages in undergraduate school. The difference is evident mostly for the A or A+ category with small increases in the proportion of students within this grade range. There is a comparable decrease in the percentage of students in the lower grade categories. The differences can probably be attributed to selection procedures of graduate schools, which enroll students with previously high performance levels, and to the "self-selection" of poorer students who do not apply to the graduate schools.

SUMMARY

The substantial majority of science students from liberal arts institutions have completed graduate studies programs. Percentages decrease for each graduating class during the period 1958–1967. However, the number of students currently enrolled increases, suggesting that the high level of graduate degree productivity will be maintained. In the specialty areas, medicine is chosen most frequently in postgraduate studies. Law attracts a small but notable ratio of graduates each year. The relative number of students enrolling in nursing (R.N.) programs or divinity studies is extremely small.

Large numbers of students change their area of specialization upon entering master's and doctorate programs. Of those whose interests are altered, about half choose major studies that are directly related to undergraduate studies. The remaining graduates choose areas that, in substance, do not reflect their undergraduate training.

One may conclude that the liberal arts undergraduate science programs function extremely well as a source of candidates for graduate school. Areas of postgraduate specialization are diverse, perhaps as a result of the undergraduate curricula variation and flexibility. Indeed, flexibility and broadness of the liberal arts education were indicated by students to be major influences in their choice of a liberal arts institution (see Chapter 5). It is plausible, then, to assume that the majority of students entering liberal arts schools are not altogether sure about their choice of a major field of study, and wish to insure a broad enough undergraduate background to be able to transfer to a new field of graduate work. Changes in postgraduate specialization do suggest the need for greater awareness of the decrease in percentage of candidates in the physical sciences. The implied risks and benefits to the student and to the institution in providing physical science training to individuals who will eventually choose another specialty also warrant some examination. Assessment of costs and benefits to both student and institution is warranted if these criteria are relevant to particular departmental policy.

Grade-point averages of the graduate students are generally higher than averages of the undergraduate major coursework performance levels. Sex differentials are most evident from data on female graduates of co-ed institutions. The women appear to perform at higher levels than the other groups considered, in both undergraduate and graduate school.

12

Financial Support for
Graduate School

Financial support for graduate studies is, of course, related to
undergraduate grades and G.R.E. scores. These indices generally
function as a basis for awarding fellowships and assistantships
to students in all institutions. Later graduate school achievement
is frequently a criterion for awarding assistantships or renewing
previous stipends.

STIPENDS

Information on the extent and source of financial stipends to
liberal arts science graduates is given in Tables 12.1, 12.2, and
12.3. The data are relevant to all graduates who have completed
advanced degrees or who are currently engaged in graduate work.
It, therefore, includes a wide variety of age groups and academic
disciplines. Stipend information was solicited using a question that
required the respondent to provide a single "best description" of
the type of stipend held: teaching assistantship, research assistantship, work-free tuition payments, work-free tuition plus expenses
payment, and no stipend support at all.

Considering Table 12.1, it is observed that between 60 and 70%
of the men who have had graduate training in each of the years
1958 to 1967 have received some form of stipend during their
course of study. The average percentage per year who state that
they have received no aid is somewhat higher for graduates of
men's colleges (39.4%) in contrast to the men from co-ed institutions (34.5%).

Of each of the response categories examined here, the work-free (tuition and grant) award appears to be most subject to consistent increase over the years. The proportion of men from co-ed institutions who receive tuition and financial supplements increased from 15% in 1958 to 23% in 1967. A similar rise is evident for graduates of the men's colleges but the proportions of recipients vary much more from year to year.

The percentage of male graduates who acknowledge the receipt of research assistantships has declined somewhat. Considering the proportions of male graduates who have had teaching assistantships, it appears that there has also been an overall decline. It is reasonable to suspect that the decrement is a function of rather stable manpower needs at institutional research facilities. That is, the number of graduate students per year may increase, but the number of assistants required for laboratory work each year may be consistent.

As shown in Table 12.2, the average yearly portion of women graduates who receive no financial aid for graduate studies is practically the same for graduates of co-ed and of women's institutions (34.5 and 31.3%, respectively). Except for certain years, the proportions generally vary moderately about these averages. No overall decrement or increment is evident for the 1958–1967 graduates. It appears that the proportion of women receiving some form of aid, without requirements for work, has increased. That is, the most evident increments in percentages are associated with the categories of work-free grants for tuition alone or for tuition augmented by other financial aid. Although the percentages vary from one graduating class to another, there is about a 7% increase in the proportion of female students receiving grants and tuition payments over the 1958–1967 period. The data on research and teaching assistantships for women fluctuate markedly, depending on the particular graduating class, and reflect no clear-cut increment or decrement. The annual fluctuations in proportions are dependent on smaller numbers of women under consideration.

	1958	1959	1960	1961
Male Co-ed	$N = 465$	$N = 504$	$N = 570$	$N = 647$
Invalid	2.9%	2.4%	3.3%	2.0%
Teaching assistantships	14.8	17.9	15.5	15.1
Research assistantships	22.5	17.5	18.0	17.6
Work-free (tuition)	8.4	10.7	9.1	9.5
Work-free (tuition + grant)	15.1	20.0	16.5	16.8
None	36.3	31.5	37.6	39.0
All-Male	$N = 247$	$N = 266$	$N = 293$	$N = 287$
Invalid	2.5%	2.3%	1.7%	2.4%
Teaching assistantships	16.3	16.2	15.4	14.0
Research assistantships	15.4	14.6	17.4	19.9
Work-free (tuition)	12.2	13.5	14.1	10.5
Work-free (tuition + grant)	10.9	12.4	11.9	16.7
None	42.7	41.0	39.5	36.5

	1958	1959	1960	1961
Female Co-ed	$N = 108$	$N = 112$	$N = 127$	$N = 174$
Invalid	0.9%	4.5%	2.3%	0.5%
Teaching assistantships	18.5	13.4	22.0	20.7
Research assistantships	10.2	8.9	7.8	13.8
Work-free (tuition)	7.4	10.7	9.4	12.7
Work-free (tuition + grant)	22.2	22.4	23.6	16.1
None	40.8	40.1	34.9	36.2
All-Female	$N = 82$	$N = 70$	$N = 75$	$N = 93$
Invalid	3.6%	1.5%	1.4%	2.1%
Teaching assistantships	19.5	11.4	20.0	17.2
Research assistantships	7.3	14.3	8.0	12.9
Work-free (tuition)	9.7	14.3	10.6	19.4
Work-free (tuition + grant)	20.8	24.3	18.7	23.7
None	39.1	34.2	41.3	24.7

TABLE 12.1

Stipends for Men's Graduate School Expenses,
by Year of Graduation

1962	1963	1964	1965	1966	1967
N = 659	N = 710	N = 733	N = 755	N = 682	N = 701
1.9%	1.4%	1.2%	0.5%	1.3%	0.5%
14.4	12.9	17.1	15.2	19.5	20.8
18.6	20.9	20.8	16.2	16.7	11.9
11.3	10.5	9.9	10.8	11.0	11.9
18.2	19.5	20.4	20.2	21.3	22.7
35.6	34.8	30.6	37.1	30.2	32.2
N = 274	N = 292	N = 357	N = 320	N = 324	N = 320
3.2%	1.1%	1.4%	1.0%	0.3%	1.5%
14.6	13.7	12.3	13.4	12.9	16.2
18.0	16.8	14.3	11.3	10.5	9.8
11.0	11.3	11.3	12.4	14.8	15.0
15.3	17.6	23.0	20.3	22.9	18.2
37.9	39.5	37.7	41.6	38.6	39.3

TABLE 12.2

Stipends for Women's Graduate School Expenses,
by Year of Graduation

1962	1963	1964	1965	1966	1967
N = 213	N = 209	N = 169	N = 206	N = 183	N = 174
1.9%	1.4%	2.4%	1.4%	2.1%	0.0%
16.5	19.6	24.2	15.5	17.0	18.9
9.0	9.0	7.7	10.7	11.5	10.4
7.5	12.9	13.6	12.6	13.7	8.6
18.8	25.4	23.1	31.5	29.5	29.9
46.3	31.7	29.0	28.3	26.2	32.2
N = 98	N = 114	N = 103	N = 108	N = 98	N = 74
4.1%	0.9%	1.0%	0.9%	1.0%	1.4%
18.3	20.1	17.5	17.6	23.4	23.0
12.2	10.6	5.9	13.9	8.2	4.0
9.2	8.8	11.6	15.8	12.3	18.9
14.3	28.1	31.0	24.1	31.6	36.5
41.9	31.5	33.0	27.7	23.5	16.2

Number of Respondents		Average (Per Year)	1958 N = 857 %	1959 N = 901 %	1960 N = 1025 %
Atomic Energy	Denied		0.0	0.1	0.0
Commission	Accepted	1.6	3.0	2.3	1.8
Department of	Denied		0.1	0.1	0.0
Defense	Accepted	1.9	2.3	2.3	2.8
National Science	Denied		6.8	5.7	6.0
Foundation	Accepted	12.1	14.0	15.1	13.4
Veteran's	Denied		0.0	0.3	0.0
Administration	Accepted	2.7	6.2	5.4	4.2
National Defense	Denied		0.0	0.6	0.0
and Education Act	Accepted	7.5	3.5	6.9	5.6
National Insti-	Denied		0.9	1.6	0.8
tutes of Health	Accepted	5.7	7.1	9.2	5.8
Fellowship Program					
National Insti-	Denied		0.3	0.3	0.1
tutes of Health	Accepted	6.8	5.3	6.5	7.8
Trainee Program					
Woodrow Wilson	Denied		1.6	3.0	3.2
Scholarship	Accepted	1.3	2.3	1.9	1.9
Other private	Denied		0.3	0.6	0.7
foundations	Accepted	8.8	9.4	11.5	10.8
Institution sources	Denied	1.2	1.0	1.0	1.0
(graduate school, etc.)	Accepted	39.4	35.8	38.7	37.5

Specifically, more extreme changes in percentage of the total sample of women is small relative to the sample for men.

Data on a variety of institutional source agencies for graduate student financial support are presented in Table 12.3. The chart contains statistics on students who applied for and were denied stipends, and the percentage of graduates who were awarded stipends, tabulated for each graduating class.

TABLE 12.3

Graduates Awarded or Denied Graduate Financial Aid,
by Year of Graduation

1961 = 1157 %	1962 N = 1173 %	1963 N = 1280 %	1964 N = 1318 %	1965 N = 1351 %	1966 N = 1250 %	1967 N = 1259 %
0.1	0.0	0.0	0.0	0.0	0.0	0.0
2.1	1.5	1.3	1.6	0.9	1.0	0.8
0.1	0.1	0.1	0.3	0.3	0.3	0.2
2.7	2.8	1.9	1.5	0.7	1.3	0.9
6.3	8.0	9.4	7.1	7.0	8.0	9.6
14.7	14.1	10.8	10.7	11.4	9.5	7.4
0.0	0.1	0.0	0.1	0.0	0.0	0.0
2.2	3.0	1.8	1.6	1.0	1.0	0.5
0.0	0.0	0.0	0.0	0.0	0.0	0.0
8.9	8.7	6.2	8.7	8.5	9.6	8.3
1.1	0.5	0.8	1.5	0.9	0.6	0.3
7.0	6.4	6.2	5.2	4.8	3.3	1.9
0.0	0.1	0.1	0.1	0.0	0.1	0.0
6.9	6.6	8.1	8.1	7.2	7.0	4.8
3.4	3.1	2.6	4.0	4.1	3.9	7.2
1.1	1.1	1.3	1.5	0.7	0.4	0.8
0.4	0.4	5.4	1.0	0.9	0.4	1.5
10.2	8.8	9.6	8.6	7.2	7.3	4.5
1.4	0.7	1.6	1.5	1.3	1.8	0.9
38.2	40.5	41.4	41.8	39.5	41.4	38.9

In terms of frequency of awards, the graduate college itself contributes most frequently to the graduate students. The percentage of graduates who obtained aid through the institution rather than directly through application to an agency ranges between 36 and 41%. Institutional sources include grants made by public or private concerns to departments or individual faculty members for maintenance or expansion and research.

The National Science Foundation (N.S.F.) programs supply financial aid to fewer students and the denial rate is much higher. However, this source ranks second in frequency with which it supplies graduate students with aid. Data for more recent years suggest that smaller proportions of students are receiving financial support from this source. If the amount of grants has been constant during the 1958–1967 period, one can infer that the total number of grants available from the N.S.F. has not kept pace with increases in graduate school enrollment.

These same inferences might be made from data for private foundation grants. Proportions are high for early years and appear to decrease slightly. The average annual fraction of students who benefit from this source of income is somewhat lower (9%) than the yearly average proportion who receive aid from the N.S.F. (12%). Private foundation sources appear to be subject to less competition insofar as denial rates are lower.

Awards made under the National Defense and Education Act (N.D.E.A.) appear to constitute a fairly stable source of graduate student funds. Except for the earliest period, the proportion of graduate students receiving such aid is maintained from year to year. The annual average proportion of awards (7.5%) coupled with the negligible rejection rate for this graduate sample suggest that applications for financial aid made to this agency are rather likely to meet with success. In rank order of proportion of stipends awarded, the National Institutes of Health (N.I.H.) fellowship or traineeship programs follow the agencies already mentioned. The annual average proportion of graduates participating in traineeship programs is 6.8%; those receiving N.I.H. fellowships account for an average of 5.7%. Both programs combined have provided partial financial support to 5–15% of the graduates, depending on year of graduation.

Three sources of funding account for small but notable numbers of students who have obtained financial aid. The Atomic Energy

Commission (A.E.C.) has provided an average of 1.6% of graduates each year with support for studies. The overall proportion of students to whom awards are made appears to be declining. Department of Defense support for graduate studies appears also to have dropped during 1958–1967. The proportion of more recent graduating classes who have received grants, and the number of students receiving aid has declined, although the sizes of graduating classes have increased.

As one would expect, the number of Veteran's Administration grants for tuition and/or educational expenses has decreased. The average yearly proportion of students who rely on this form of aid is slightly larger (2.7%) than that of Department of Defense programs or A.E.C. awards, and has decreased more quickly. The extent to which this decrement is a result of decreasing number of Korean War veterans, independent of those students who have served in Viet Nam, is not estimable from these data.

Relatively high rates of rejection compared to the proportion of acceptances are evident for Woodrow Wilson Fellowships. The declining proportion in awards is evident largely for the three most recent graduating classes. Until 1965 it appears that the rate of awards matched increases in the number of graduate students.

It appears that the proportion of financial support from the agencies under examination has declined and decrements are larger or smaller depending on the specific source. In terms of rank order of frequency of award, institutional sources, the N.S.F., private foundations and the N.D.E.A. furnish the largest numbers of students with support annually. More substantial decrements in proportion of students receiving awards from 1958–1967 are evident from the data on the A.E.C., Department of Defense, N.I.H. fellowship programs, and Veteran's Administration support. Insofar as maintaining a constant percentage of awards (rather than a constant number of them), the other agencies appear to be keeping closer pace with graduate school enrollments.

SUMMARY AND DISCUSSION

Financial support for graduate schools appears to be increasing in the direction of work-free awards, rather than teaching or research assistantship programs. The proportion of students who receive either work-free or work-study stipends is high, accounting for more than 60% of the graduates. Considering specific funding agencies, the awards are most frequently derived from institutional sources (e.g., the graduate college), N.S.F., N.D.E.A. funds and from private foundations. The rate of increase in graduate school enrollment appears to be higher than the rate at which stipend distribution can be increased. That is, the proportion of graduates who are supported by outside agency funds has declined somewhat during 1958–1967.

The statistics on financial aid imply that the proportion of graduate students who are required for research or teaching assistant functions is decreasing. Work-free awards, on the other hand, are being awarded more frequently. An overall increase in percentage of graduates receiving assistantship or fellowship awards from major public and private agencies may be indicative of future increased competitivenss among entering graduate students. The decrease in aid may also lead to decreases in the rate of graduate school enrollments.

13

Postgraduate Employment

The extent and diversity of full-time work following graduation is interesting in several respects. One might predict that men in this sample do not generally enter the labor force until well after receipt of their baccalaureate degree, and also that their chosen occupations are differentially related to major coursework depending on sex, major field, and so forth. The same inference may be made about women's occupational development. Data in this chapter were obtained in order to more accurately define the similarities and differences of the graduates across each of the science groups for the 1958–1967 period. In addition to duration of full-time employment, the nature of the employment is also examined.

DURATION OF EMPLOYMENT

A survey was made to determine the percentage of graduates in each graduating class who were gainfully employed on a full-time basis in each succeeding year from 1958 to 1967. Tables 13.1 and 13.2 are tabulations of these data for men and women, respectively.

Largely because of the high rates of graduate school attendance, one can expect that duration of full-time employment in the sciences is markedly low during 1958–1967. The majority of graduating males defer full-time employment until four years after completion of their undergraduate educations (Table 13.1). The majority of graduates who are not officially employed on a full-time basis are in graduate schools. Fewer individuals are likely to

	1958 N = 870 %	1959 N = 933 %	1960 N = 1014 %	1961 N = 1121 %	1962 N = 1120 %
In 1958	13.8	0.9	1.0	1.2	0.1
In 1959	26.6	16.5	2.0	1.3	0.4
In 1960	32.5	28.0	17.0	1.2	0.7
In 1961	37.2	32.2	26.3	17.9	1.1
In 1962	48.5	38.3	30.0	27.9	18.8
In 1963	57.8	48.2	35.0	32.7	26.5
In 1964	65.4	58.0	46.0	37.6	31.0
In 1965	70.1	65.3	57.8	46.4	35.7
In 1966	74.0	72.7	65.2	58.3	45.5
In 1967	77.8	75.9	69.3	65.7	54.6

	1958 N = 365 %	1959 N = 357 %	1960 N = 405 %	1961 N = 468 %	1962 N = 471 %
In 1958	45.5	0.8	0.3	0.4	0.2
In 1959	63.6	44.5	0.7	0.0	0.2
In 1960	61.4	65.6	48.4	0.2	0.2
In 1961	51.8	59.9	67.2	39.5	1.1
In 1962	42.2	52.9	62.7	59.2	41.0
In 1963	40.3	47.9	54.6	59.2	62.2
In 1964	35.6	42.9	45.2	54.5	63.9
In 1965	33.2	38.1	44.0	48.7	56.1
In 1966	35.1	35.3	36.8	44.7	51.6
In 1967	34.8	34.2	34.8	40.6	50.7

TABLE 13.1

		Full-Time Employment of Male Graduates, by Year of Graduation		
1963 N = 1181 %	1964 N = 1285 %	1965 N = 1338 %	1966 N = 1292 %	1967 N = 1363 %
1.0	0.8	1.0	0.7	0.2
1.2	0.9	1.0	0.8	0.2
1.0	1.0	1.3	0.9	0.2
0.8	1.0	1.4	1.1	0.4
0.9	1.1	1.5	1.5	0.8
17.3	1.3	1.5	1.2	1.3
24.9	13.9	1.8	1.3	1.3
30.3	22.5	13.5	1.6	1.3
34.5	26.8	22.5	12.5	1.1
43.7	31.1	28.7	19.7	13.0

TABLE 13.2

		Full-Time Employment of Female Graduates, by Year of Graduation		
1963 N = 543 %	1964 N = 515 %	1965 N = 579 %	1966 N = 562 %	1967 N = 579 %
0.2	3.5	0.0	0.0	0.4
0.2	0.2	1.0	0.5	0.2
0.6	0.4	0.2	0.2	0.2
0.4	0.6	0.2	0.4	0.2
0.4	0.4	0.2	0.2	0.4
44.0	1.2	0.2	0.2	0.4
62.6	45.4	1.0	0.4	0.4
63.7	62.3	39.2	0.9	0.5
61.3	67.6	61.0	43.2	0.9
56.7	65.1	65.6	61.7	39.2

be in temporary military services, engaged in part-time consulting work, or other similar activities. In the case of male graduates, there appears to be a slight tendency for fewer individuals to start work immediately following graduation in more recent years. This differential is likely to be a function of slight increases in the rate of graduate school enrollments and, also, of Federal Selective Service criteria and quota variations.

The pattern for female graduate employment is different from that of the men (Table 13.2). Larger numbers enter the labor force soon after graduation, though there appears to be a trend in the direction of smaller percentages for more recent classes. The majority of women have full-time jobs two or three years after graduation, as opposed to four years for males. The ratio of women employed to total class size decreased markedly four years after graduation. About 35% of the graduates remain in the labor force after their colleagues have discontinued full-time work. Data presented earlier on graduate school attendance (Chapter 11) provide some evidence that the reduction in full-time employment is not attributable entirely to marriage. Women enter graduate schools somewhat later than men, and this phenomenon contributes to the attrition from full-time employment categories.

NATURE OF EMPLOYMENT

Some rather broad categories can be specified in order to assay the nature of graduate employment. Data are presented in Tables 13.3 and 13.4 by sex of the graduate, for thirteen major types of commercial and professional occupations. Graduates were also requested to specify whether they had joined a first, second, third or fourth employer.

The major percentage of graduates enter private industry or business, the single largest category of response for men. Business also attracts women but the relative percentages are about one-fifth less than corresponding data for men. A strong competitor for women's skills are secondary school systems. From 20 to 25%

of female graduates are employed in administrative or teaching functions at this level. The consistent ratio of women entering this field, regardless of the sequence of employers, suggests that a very stable pool of job candidates for school staffs is being provided by the liberal arts college.

The second major occupation for men is the university or college environment (nonmedical). Teaching, research or administrative positions account for about 15% of the male graduates' first and second employers. The slight drop in magnitude of percentages for third and fourth employers may be attributable to decreasing interest in academic work, at least for individuals who are likely to change employers frequently after graduation.

The data for men are similar to the information about women's employers insofar as teaching in the primary and secondary school systems receives notable attention from both groups. Approximately 10% of the men choose staff work in this area.

Women are more likely than men to choose a medical school environment or nonprofit hospitals, clinics and agencies as employers.

State and local governments attract very few science graduates, but a fairly stable percentage (5%) of men and women appear to choose the Federal government as an employer. Military service is a first or second employer of men, but the ratio decreases markedly for third and fourth employer choices.

Self-employment is most frequently affirmed by graduates who have had at least three previous jobs. Both men and women follow this pattern, but the proportion of women is only one-third that of men who are self-employed.

OCCUPATIONAL TYPES

The first full-time position of the science graduates reflects a number of important variables considered earlier. The specific major discipline, the type of graduate or professional school training, sex, and type of employer are the more obvious determinants of

Employer

Invalid
Private industry or business
Self-employed
College or university
   (other than medical)
Secondary school or school system
Medical school
Federal government—civilian employee
U.S. Public Health Service
   Commissioned Corps
Military service—active duty
State government
Local government
Nonprofit hospital or clinic
Other nonprofit organization
Other
Total

the first occupational role chosen by the graduate. The graduate questionnaire (Appendix B3) permits respondents to designate one of 43 occupational types as a first, full-time position. Statistical data are presented in Table 13.5 with respondent information labelled according to field of study and sex.

Consider the biology graduates. Very few obtain first, full-time jobs as biologists. Indeed, secondary school teaching attracts a higher proportion of biology graduates (15%) than any other occupation. A substantial number of the graduates become physicians or surgeons after completion of biological or professional training. About 13% become laboratory technicians or hygienists. These

TABLE 13.3

First Employer	Second Employer	Male Graduates Employed by Various Types of Employers Third Employer	Fourth Employer
N = 6300	N = 2588	N = 944	N = 377
%	%	%	%
1.3	0.6	0.8	0.8
40.5	40.9	45.3	43.3
3.4	4.6	4.6	9.0
14.7	15.2	11.6	9.0
11.1	10.2	9.1	11.9
4.3	2.6	4.5	4.5
5.7	4.7	5.2	4.8
0.7	1.1	0.7	0.5
7.6	8.9	4.9	3.7
1.3	1.5	2.1	1.1
0.8	0.9	1.4	1.3
5.2	4.7	4.8	6.1
2.2	2.7	3.3	2.9
1.3	1.0	1.5	1.1
100.1	99.6	99.8	100.0

occupations (each of which account for 10–15% of the employed biology graduates) account for nearly 50% of employed graduates who majored in biology.

A second group involves occupational categories that comprise 3 to 10% of the biology graduates. Medical or dental technicians include 5.7%. Engineering or science technicians, dentists, and college teachers together include about 10% of the biology graduates. It is interesting to note that, although a substantial number of the graduates entered business, there are few commercial salesmen or executives from the biology ranks, or from any other of the science classifications for that matter.

Employer

Invalid
Private industry or business
Self-employed
College or university
   (other than medical)
Secondary school or school system
Medical school
Federal government—civilian employee
U.S. Public Health Service
   Commissioned Corps
Military service—active duty
State government
Local government
Nonprofit hospital or clinic
Other nonprofit organization
Other
Total

The chemistry graduates are similar to those from the pre-med programs insofar as the largest segment enters a closely related field. Some 42% of the chemistry majors actually state that their full-time occupation is chemist. Medical schools capitalize on part of the chemistry pool: nearly 8% of the graduates said that their first full-time positions were as physicians or surgeons. Laboratory technical work and engineering science technician jobs account for equal percentages of graduates. The only remaining activity which includes a notable ratio of chemistry graduates is secondary school teaching. Far fewer chemists enter this field than biologists.

Of the pre-med students who did not become physicians (60%

TABLE 13.4

		Female Graduates Employed by Various Types of Employers	
First Employer N = 3861 %	Second Employer N = 1946 %	Third Employer N = 810 %	Fourth Employer N = 331 %
1.1	1.2	0.7	1.5
29.8	24.2	24.1	23.5
0.5	1.7	2.0	3.6
14.7	17.0	14.6	13.0
22.3	23.6	23.0	25.6
8.1	7.9	7.3	4.8
5.0	5.1	4.8	4.5
0.1	0.1	0.1	
0.2	0.1	0.1	
1.5	1.2	2.7	2.4
0.9	1.3	1.5	2.4
9.6	9.3	10.1	7.9
4.2	4.8	4.7	4.8
2.0	5.0	4.3	5.7
100.0	100.1	100.0	99.7

of the total), a substantial number of individuals entered dentistry (8.2%) and secondary school teaching (7.5%). These two activities attract larger proportions of pre-med graduates than any other remaining occupations. A tertiary grouping can be constructed from the technical assistance categories. An additional 5.2% of the pre-med graduates became laboratory technicians.

A substantial number of mathematics graduates had a first full-time position as a computer programmer. An almost equal number entered secondary schools as teachers. The remaining occupational classifications each include less than one-fifth the total percentage in these two professions. Of the remaining groups,

	Male N = 6402 %	Female N = 3980 %
Invalid	3.9	3.1
Accountant or actuary	2.5	0.8
Actor or entertainer	0.0	0.0
Architect	0.1	0.1
Artist	0.0	0.2
Athlete or recreation director	0.2	0.2
Business executive (management, administration)	4.0	0.5
Business owner or proprietor	0.4	0.0
Business salesman or buyer	2.8	0.4
Clergyman (minister, priest)	0.4	0.0
Clergy (other religious)	0.1	0.0
Clerical worker	0.7	2.6
Clinical psychologist	0.1	0.0
College teacher	7.3	3.0
Computer programmer	5.2	10.6
Conservationist or forester	0.1	0.0
Dentist (including orthodontist)	2.6	0.0
Designer or draftsman	0.2	0.0
Dietitian or home economist	0.0	0.0
Engineers		
Aeronautical	0.5	0.0
Civil	0.5	0.0
Electrical	2.4	0.1
Mechanical	1.1	0.0
Sales	0.8	0.0
Other	3.8	1.0
Farmer or rancher	0.2	0.0
Foreign service worker (incl. diplomat)	0.1	0.0
Housewife	0.0	5.6
Interior decorator (incl. designer)	0.0	0.0
Interpreter (translater)	0.0	0.0
Lab technician or hygienist	1.5	11.4
Law enforcement officer	0.1	0.0
Lawyer (attorney or judge)	0.8	0.1

TABLE 13.5

First Full-Time Occupations of Graduates,
by Sex and by Major Subject

Biology N = 3502 %	Chemistry N = 1980 %	Math N = 2943 %	Physics N = 1425 %	Pre-med N = 536 %
4.4	3.2	3.0	3.9	2.0
0.2	0.3	5.7	0.6	0.9
0.1	0.0	0.0	0.0	0.0
0.0	0.0	0.1	0.1	0.2
0.2	0.0	0.0	0.0	0.0
0.3	0.1	0.1	0.1	0.4
2.2	1.4	3.9	3.4	1.5
0.2	0.3	0.2	0.3	0.4
2.9	1.8	1.2	0.9	1.9
0.3	0.1	0.2	0.1	0.4
0.0	0.1	0.0	0.1	0.0
1.6	0.6	2.3	0.3	1.5
0.1	0.1	0.0	0.0	0.0
3.9	7.1	6.6	7.8	1.3
0.5	1.0	21.7	5.5	0.2
0.2	0.0	0.0	0.1	0.0
3.2	0.6	0.1	0.0	8.2
0.0	0.0	0.2	0.5	0.0
0.0	0.0	0.0	0.0	0.0
0.0	0.0	0.4	1.1	0.2
0.0	0.1	0.5	1.2	0.0
0.0	0.1	0.5	9.8	0.0
0.0	0.1	0.5	3.4	0.2
0.0	1.0	0.5	0.9	0.0
0.1	2.3	3.5	9.1	0.4
0.2	0.2	0.0	0.1	0.4
0.1	0.0	0.1	0.1	0.0
3.0	1.7	2.3	0.6	1.5
0.0	0.0	0.0	0.1	0.0
0.0	0.0	0.1	0.0	0.0
12.7	4.5	0.3	0.2	5.2
0.2	0.0	0.0	0.1	0.0
0.4	0.4	0.8	0.6	0.9

(continued)

	Male N = 6402 %	Female N = 3980 %
Military service (career)	3.3	0.1
Musician (performer, composer)	0.0	0.0
Natural scientists		
Biologist	3.4	5.4
Chemist	10.2	6.6
Geologist	0.1	0.0
Mathematician	1.7	1.8
Physicist	5.8	0.9
Nurse	0.1	0.6
Optometrist	0.1	0.0
Personnel or labor relations	0.1	0.2
Pilot	0.4	0.0
Pharmacist	0.1	0.0
Radio or T.V.	0.1	0.1
Physician or surgeon	10.9	2.0
Real estate broker	0.1	0.0
School counselor	0.1	0.1
School principal or superintendent	0.0	0.0
Social worker	0.4	0.7
Statistician	0.7	1.2
Therapist (physical, occupational, speech)	0.1	0.9
Teacher, elementary	0.5	2.4
Teacher, secondary	10.5	19.2
Technicians		
Medical or dental	0.6	5.1
Electrical/electronic	0.3	0.0
Engineering/science	1.2	4.3
Veterinarian	0.2	0.0
Writer or journalist	0.3	0.7
Skilled trades	0.3	0.1
Other	6.8	8.1

TABLE 13.5 (continued)

Biology N = 3502 %	Chemistry N = 1980 %	Math N = 2943 %	Physics N = 1425 %	Pre-med N = 536 %
1.3	2.1	2.3	3.5	1.5
0.0	0.0	0.0	0.1	0.0
10.2	1.9	0.2	0.6	4.7
1.5	42.1	0.2	0.4	2.8
0.0	0.1	0.1	0.1	0.0
0.0	0.0	5.7	0.8	0.0
0.0	0.5	0.9	25.8	0.0
0.6	0.0	0.0	0.1	0.2
0.1	0.0	0.0	0.0	0.2
0.2	0.0	0.1	0.1	0.4
0.3	0.1	0.2	0.5	0.0
0.1	0.0	0.0	0.0	0.2
0.1	0.0	0.1	0.1	0.2
11.4	7.5	0.2	0.4	41.0
0.0	0.0	0.1	0.0	0.2
0.1	0.0	0.1	0.0	0.0
0.0	0.0	0.0	0.0	0.0
0.9	0.2	0.3	0.1	0.2
0.0	0.1	2.8	0.4	0.0
1.0	0.0	0.0	0.0	0.6
2.2	0.3	1.0	0.3	1.3
15.2	6.9	21.4	6.9	7.5
5.7	1.1	0.0	0.0	3.7
0.0	0.1	0.2	0.9	0.0
3.0	3.5	1.0	2.5	2.0
0.3	0.2	0.0	0.0	0.4
0.6	0.4	0.5	0.4	0.6
0.4	0.1	0.1	0.4	0.0
8.1	5.6	8.7	5.7	4.8

the most likely full-time positions to be chosen are accounting, mathematics or college teaching, or the engineering disciplines. A bit more than one-quarter of the physics major group actually became physicists following their education. Additional smaller numbers are engaged in college teaching. The physics departments contribute strongly to engineering manpower (25% of the physics graduates). Physics majors are more likely to enter an engineering field than are graduates from any other of the science programs. Computer programming and secondary school teaching attract many of the remaining students. The physics majors are similar to those from mathematics programs insofar as very few of them take full-time positions in the technical assistance areas.

Very small but consistent numbers of graduates from each of the major discipline categories enter a wide variety of other occupations. From 0.5% to 1.0% of the graduates within each of the disciplines considered enter journalistic or writing occupations. About the same percentage of individuals become school counselors or social workers. From 0.4 to 0.9% of each category indicated that their first full-time position was as an attorney.

Female graduates are most likely to be employed as secondary school teachers (19.2%), lab technicians (11.4%), or computer programmers (10.6%). Male graduates are more frequently first employed as natural scientists (21.2%), physicians (10.9%), secondary school teachers (10.5%) or engineers (9.1%).

SUMMARY

The majority of men in this sample do not engage in full-time employment until nearly four years after graduation, while women enter the full-time labor force about a year sooner. Most of the female science graduates were employed full-time within two to three years after completing undergraduate education. The delay for men is largely attributable to graduate school attendance. Since many women do not attend graduate school until later in their careers, the immediate impact of this factor on the labor pool of women is attenuated. Most men enter commercial business enterprises as a first full-time position. Both business and secondary schools account for a substantial ratio of female graduates. Although their employment is frequently in a business environment, relatively few graduates take jobs as commercial executives or salesmen. A wide range of occupations is evident, and many of the first-time employment categories are only tangentially related to undergraduate training. Secondary school teaching is an exceedingly attractive occupation for most of the science groups independent of sex. Comparing across major studies categories, it appears that engineering work is most likely to attract the physics graduates, whereas computer programming is most attractive to mathematics majors. Chemistry and pre-med students most frequently adhered to their respective major areas of interest upon entering the labor force.

14

Postgraduate Professional Achievement

The variety and extent of postgraduate professional activities are documented in this chapter. Science graduates provided information on the number of articles or books published and the number of papers presented at professional meetings. Data on inventions, awards, and membership in professional associations were also compiled. Such activities naturally reflect the development of expertise and professional stature, and can be considered as a multifaceted index of success in the graduate's field. Insofar as many graduates are employed in occupations which are not directly related to science, the professional activities considered here include nontechnical as well as scientific enterprises.

PUBLICATION OF BOOKS, ARTICLES, AND PAPERS

The earliest class contains the largest proportion of individuals who have written books (Tables 14.1 and 14.2). Nearly 2% of the male graduate sample, and about 1.5% of the female sample acknowledged writing one or more books. This ratio attenuates for the ensuing classes to about 1% of the men's group up to 1963. The fraction of women decreases dramatically from 1% in 1958 to nearly 0% in 1963. It is evident that the men are more productive than women with respect to this activity and also complete books before women do.

In the more recent classes, some interest in writing books is evident. A sharp decline in the ratio of men who have written at least one book occurs between 1964 and 1967: 0.6%, 0.4%, 0.2%, and 0.1%, for each respective year. In the women's sample a similar, low plateau of activity averages to about 0.4% of the total group for the same period.

The number of graduates who acknowledge publication of articles is somewhat more substantial. Table 14.3 contains available statistics based on the male respondents from each graduating class. Similar data on women's productivity are given in Table 14.4.

For the men, there appears to be a fairly consistent ratio (10 to 12%) of individuals from each graduating class between 1958 and 1964 who have written a single article. Fewer men responded that they had published two or three articles during this same period, and the ratio varies between 1.6 and 14.2% of the total sample. Consistently productive writers are evident in the response category of four or more publications. The 1958, 1959, and 1960 classes have high proportions (14.5 to 16.9%) of their respective graduates at this level of activity. Again, the ratio of respondents in this category decreases markedly as one considers the more recent graduating classes. The influence of age, and perhaps persistence, is evident in the increasing proportions of men who acknowledged that they had undertaken the activity but had not succeeded in publishing.

The data on female graduates indicate, as one might expect, that they publish fewer articles than men. In the 1958–1964 classes, the proportion of women who said that they had published one article fluctuates more than the corresponding ratio of men, and is 3 to 4% lower than the statistics for men during that period. The decrement in proportion of women with one publication is steeper for 1965–1967 than the corresponding figures for men.

Statistics for women with higher publication rates are also typically below the men's level, but a consistent number of active women is evident in each graduating class. From 5 to 8% of the female graduates of 1958–1964 have written two or three articles,

	1958 *N = 870* %	*1959* *N = 933* %	*1960* *N = 1014* %	*1961* *N = 1121* %
Invalid	0.2	0.2	0.2	0.3
No response	18.7	19.9	19.5	17.9
One	1.6	0.8	0.7	0.8
Two	0.1	0.0	0.2	0.0
Three	0.0	0.2	0.2	0.0
Four or more	0.1	0.1	0.0	0.1
Not attempted	75.8	76.3	75.9	79.1
Undertaken, but not achieved	3.5	2.5	3.3	1.8

	1958 *N = 365* %	*1959* *N = 357* %	*1960* *N = 405* %	*1961* *N = 468* %
Invalid	0.6	0.0	0.0	0.0
No response	22.7	23.0	20.5	19.9
One	0.8	0.8	0.7	0.9
Two	0.3	0.0	0.3	0.0
Three	0.0	0.3	0.0	0.0
Four or more	0.3	0.0	0.0	0.2
Not attempted	73.4	74.2	76.8	78.2
Undertaken, but not achieved	1.9	1.7	1.7	0.9

TABLE 14.1

				Books Written by Male Graduates, by Year of Graduation	
1962 *N = 1120* %	*1963* *N = 1181* %	*1964* *N = 1285* %	*1965* *N = 1338* %	*1966* *N = 1292* %	*1967* *N = 1363* %
0.0	0.1	0.0	0.1	0.1	0.0
20.7	18.5	20.0	18.4	21.2	20.8
0.8	0.9	0.5	0.3	0.2	0.1
0.2	0.0	0.1	0.0	0.0	0.0
0.1	0.1	0.0	0.1	0.0	0.0
0.0	0.0	0.0	0.0	0.0	0.0
76.1	79.3	78.4	79.8	77.9	78.4
2.1	1.0	1.0	1.4	0.7	0.8

TABLE 14.2

				Books Written by Female Graduates, by Year of Graduation	
1962 *N = 471* %	*1963* *N = 543* %	*1964* *N = 515* %	*1965* *N = 579* %	*1966* *N = 562* %	*1967* *N = 579* %
0.4	0.0	0.0	0.2	0.0	0.0
20.0	18.6	18.3	19.9	22.2	19.5
0.6	0.0	0.2	0.5	0.5	0.2
0.0	0.0	0.0	0.0	0.0	0.0
0.0	0.0	0.0	0.0	0.0	0.0
0.2	0.0	0.0	0.0	0.0	0.0
77.7	80.1	80.6	79.3	76.9	79.3
1.1	1.3	1.0	0.2	0.4	1.0

	1958 N = 870 %	1959 N = 933 %	1960 N = 1014 %	1961 N = 1121 %
Invalid	4.3	5.6	4.7	4.4
No response	7.4	9.1	8.3	7.9
One	11.6	11.2	10.2	9.5
Two	7.8	7.7	7.6	8.3
Three	4.3	4.7	5.2	5.9
Four or more	16.9	16.4	14.5	9.9
Not attempted	39.4	37.5	39.5	43.3
Undertaken, but not achieved	8.4	7.8	10.1	11.0

	1958 N = 365 %	1959 N = 357 %	1960 N = 405 %	1961 N = 468 %
Invalid	1.6	1.4	1.2	3.2
No response	19.7	14.0	15.6	12.8
One	4.9	8.4	6.7	7.5
Two	4.4	4.2	4.4	4.9
Three	2.7	2.2	3.5	3.6
Four or more	6.0	6.2	4.7	3.2
Not attempted	56.7	59.1	60.3	59.4
Undertaken, but not achieved	3.8	4.5	3.7	5.3

TABLE 14.3

Articles Published by Male Graduates,
by Year of Graduation

1962 N = 1120 %	1963 N = 1181 %	1964 N = 1285 %	1965 N = 1338 %	1966 N = 1292 %	1967 N = 1363 %
3.9	3.6	5.8	3.7	2.2	1.4
10.7	10.6	11.0	11.7	15.8	17.6
11.5	10.5	10.4	7.0	6.0	3.1
5.4	6.9	4.3	2.8	1.1	1.4
3.0	3.1	1.6	1.1	0.5	0.2
8.2	4.9	3.7	1.4	0.9	0.2
46.2	47.3	49.6	57.6	60.7	65.4
11.1	13.2	13.7	14.9	12.9	10.7

TABLE 14.4

Articles Published by Female Graduates,
by Year of Graduation

1962 N = 471 %	1963 N = 543 %	1964 N = 515 %	1965 N = 579 %	1966 N = 562 %	1967 N = 579 %
3.6	2.4	2.1	2.1	2.1	0.4
14.7	12.3	14.0	15.2	16.7	17.8
8.5	10.5	7.2	6.4	3.9	1.2
3.0	3.5	3.9	2.9	0.5	0.4
2.3	2.2	1.6	1.4	0.5	0.2
2.3	3.9	1.6	0.5	0.4	0.0
59.7	59.3	62.1	64.6	65.0	71.7
5.9	5.9	7.6	6.9	10.9	8.5

and this ratio drops to 0.6% for the most recent graduates. The numbers of highly productive women (four articles or more) vary from year to year and decrease from 6% to 0.4% for the 1958–1966 groups.

Presentation of papers at professional meetings appeals to a substantial number of men and fewer women (Tables 14.5 and 14.6). In general, the pattern and extent of men's activity is similar to that described for article publication. Within the women's group, the pattern is altered insofar as smaller proportions are evident.

Consider first the data on male graduates. From 11 to 16% of respondents from classes of 1958 through 1964 delivered one professional paper. Two papers were presented by an average of 9% of graduates from 1958 to 1961. In both categories of response, the proportions decrease sharply in later years. Nearly 6% of the 1967 graduates had presented a single paper, and only 1.5% of the graduates had delivered two or three. Individuals with a high rate of presentation (four or more papers) comprise 14% of the earliest class. This ratio decreases with fair uniformity from 12 to 0.6% for the remaining classes.

The proportion of women who have presented papers is generally much lower than corresponding statistics for men. In the earliest two classes, for example, approximately 40% of the men acknowledged delivery of at least one paper. On the other hand, only 14% of the women from this class made a similar claim. The higher productivity rates are generally associated with older graduates, as evidenced by the ratio of women who have delivered four or more papers in the 1967 sample.

INVENTIONS

Science graduates' inventions are a useful way of assessing one aspect of nonliteral productivity. The occupation is a biased index of activity insofar as chemistry or physics majors may be more likely to find inventing to be of interest, than are the mathematics and biology majors. Further, the conception of "invention" may differ for respondents, depending on their expertise in using existing materials to create new (presumably) mechanical devices or to revise old ones.

In the men's sample, an average of 1.9% of the entire group affirm that they have developed an invention (Table 14.7). About 0.4% of the women's sample make this same claim (Table 14.8). Two or more inventions are cited by an additional 2% of the male science graduates, in contrast to 0.4% of the women in the same response category.

The decrease in proportion of men who have devised at least one invention is evident through the 1958–1967 period. The 3% level appears to be a plateau for this category of response and includes 1959 and 1960 graduates also.

As might be expected, inventions by women are much less frequent than by men (Table 14.8).

RESEARCH GRANTS

Tables 14.9 and 14.10 contain data on the frequency with which science graduates acquire research grants. The data must be qualified to the extent that the physical science research funds from private or public agencies are typically larger than financial subsidy available, for example, to the natural sciences. Although chemistry and physics majors may be most likely to have been recipients of research grants, the exact proportion of pre-med, biology, or math recipients is impossible to determine without further analysis of the data.

From Table 14.9 it is apparent that an average of 10% of the men in each graduating class have applied for and received only one research grant. Between 7 and 9% of the 1958–1967 graduates have been awarded two or three grants. The ratio drops considerably (to 1.3%) for the youngest members of the sample. Small

	1958 N = 870 %	1959 N = 933 %	1960 N = 1014 %	1961 N = 1121 %
Invalid	0.3	0.6	1.1	0.7
No response	12.0	13.3	13.2	12.1
One	11.5	14.9	13.2	15.5
Two	10.0	8.0	9.2	8.1
Three	5.3	3.3	4.3	3.0
Four or more	13.9	12.0	10.5	7.6
Not attempted	44.8	44.8	46.0	49.7
Undertaken, but not achieved	2.2	3.0	2.6	3.2

	1958 N = 365 %	1959 N = 357 %	1960 N = 405 %	1961 N = 468 %
Invalid	0.3	0.3	0.0	0.0
No response	21.9	20.5	19.0	18.4
One	7.4	6.4	6.9	4.9
Two	2.2	2.0	1.7	3.0
Three	1.4	1.4	1.2	1.5
Four or more	3.0	2.5	1.7	1.9
Not attempted	63.6	65.8	67.7	68.4
Undertaken, but not achieved	0.3	1.1	1.7	1.9

TABLE 14.5

Papers Presented by Male Graduates,
By Year of Graduation

1962 N = 1120 %	1963 N = 1181 %	1964 N = 1285 %	1965 N = 1338 %	1966 N = 1292 %	1967 N = 1363 %
0.7	0.3	0.5	0.4	0.4	0.2
15.4	15.0	15.6	14.9	19.4	19.8
13.9	14.2	11.4	9.4	6.4	5.5
6.5	6.5	5.5	4.4	2.4	1.2
3.0	2.8	1.6	1.1	0.7	0.3
4.6	2.5	2.7	2.0	1.1	0.6
53.5	55.7	59.5	64.3	66.3	69.7
2.4	2.9	3.4	3.6	3.4	2.7

TABLE 14.6

Papers Presented by Female Graduates,
by Year of Graduation

1962 N = 471 %	1963 N = 543 %	1964 N = 515 %	1965 N = 579 %	1966 N = 562 %	1967 N = 570 %
0.0	0.0	0.0	0.2	0.0	0.2
19.3	16.4	17.7	18.8	20.8	19.5
5.7	9.9	4.5	3.6	4.3	1.9
1.7	2.0	1.4	1.2	1.1	0.0
0.6	0.7	0.0	0.9	0.2	0.0
1.1	0.6	0.4	0.4	0.2	0.2
70.7	68.9	74.0	72.4	71.9	76.5
0.9	1.5	2.1	2.6	1.6	1.7

	1958 N = 870 %	1959 N = 933 %	1960 N = 1014 %	1961 N = 1121 %
Invalid	0.3	0.2	0.2	0.1
No response	20.8	21.5	19.9	17.8
One	3.0	2.1	3.3	3.7
Two	1.4	1.4	1.5	1.3
Three	0.2	0.5	0.8	0.4
Four or more	1.8	1.8	2.0	1.1
Not attempted	69.9	70.7	70.8	72.9
Undertaken, but not achieved	2.5	1.6	1.6	2.8

	1958 N = 365 %	1959 N = 357 %	1960 N = 405 %	1961 N = 468 %
Invalid	0.0	0.0	0.0	0.0
No response	23.0	23.8	21.5	21.4
One	0.8	0.0	0.7	0.4
Two	0.6	0.6	0.0	0.0
Three	0.3	0.3	0.0	0.0
Four or more	0.0	0.3	0.5	0.2
Not attempted	75.3	74.8	77.3	77.8
Undertaken, but not achieved	0.0	0.3	0.0	0.2

TABLE 14.7

Inventions Made by Male Graduates,
by Year of Graduation

1962 N = 1120 %	1963 N = 1181 %	1964 N = 1285 %	1965 N = 1338 %	1966 N = 1292 %	1967 N = 1363 %
0.4	0.3	0.1	0.0	0.1	0.2
22.1	19.5	20.9	19.4	22.5	21.5
2.1	1.9	1.1	1.1	0.3	0.2
1.0	0.6	0.9	0.2	0.3	0.0
0.5	0.2	0.5	0.2	0.0	0.0
0.3	0.7	0.5	0.3	0.2	0.1
71.7	75.2	74.6	77.2	75.4	76.8
2.0	1.7	1.6	1.7	1.2	1.3

TABLE 14.8

Inventions Made by Female Graduates,
by Year of Graduation

1962 N = 471 %	1963 N = 543 %	1964 N = 515 %	1965 N = 579 %	1966 N = 562 %	1967 N = 579 %
0.0	0.0	0.0	0.2	0.0	0.0
21.7	20.3	19.0	20.2	22.8	20.4
0.0	0.4	0.8	0.4	0.5	0.0
0.0	0.0	0.0	0.4	0.0	0.2
0.0	0.0	0.0	0.0	0.0	0.0
0.0	0.2	0.0	0.0	0.0	0.0
78.1	78.8	80.0	78.4	76.2	78.2
0.2	0.4	0.2	0.5	0.5	1.2

	1958 N = 870 %	1959 N = 933 %	1960 N = 1014 %	1961 N = 1121 %
Invalid	0.0	0.4	0.2	0.2
No response	14.8	16.3	16.7	15.3
One	10.3	11.5	10.6	11.1
Two	5.1	6.2	5.0	4.6
Three	2.3	2.4	1.3	2.1
Four or more	3.7	2.3	2.0	1.0
Not attempted	62.3	59.5	61.6	64.1
Undertaken, but not achieved	1.5	1.5	2.7	1.7

	1958 N = 365 %	1959 N = 357 %	1960 N = 405 %	1961 N = 468 %
Invalid	0.0	0.0	0.3	0.0
No response	23.3	23.3	20.5	18.6
One	4.4	2.5	3.5	4.5
Two	1.1	0.8	1.7	1.9
Three	0.3	1.4	0.5	0.6
Four or more	0.8	0.8	0.3	0.2
Not attempted	69.6	70.6	72.8	73.5
Undertaken, but not achieved	0.6	0.6	0.5	0.6

TABLE 14.9

			Research Grants to Male Graduates, by Year of Graduation		
1962 *N = 1120* %	*1963* *N = 1181* %	*1964* *N = 1285* %	*1965* *N = 1338* %	*1966* *N = 1292* %	*1967* *N = 1363* %
0.2	0.4	0.5	0.6	0.2	0.7
19.6	17.0	17.2	16.1	19.3	18.9
8.4	9.2	9.4	9.9	10.4	9.7
3.7	4.3	4.9	4.0	3.3	1.1
2.1	2.1	1.6	1.2	0.6	0.2
1.2	0.6	0.9	0.2	0.5	0.1
62.9	64.6	64.3	66.8	65.1	67.9
2.0	1.7	1.2	1.1	0.6	1.5

TABLE 14.10

			Research Grants to Female Graduates, by Year of Graduation		
1962 *N = 471* %	*1963* *N = 543* %	*1964* *N = 515* %	*1965* *N = 579* %	*1966* *N = 562* %	*1967* *N = 579* %
0.0	0.4	0.0	0.0	0.4	0.0
20.2	17.9	17.9	19.2	21.0	20.0
5.5	3.3	4.5	3.8	4.6	3.8
1.9	2.6	1.4	2.1	1.4	0.5
0.6	0.4	0.6	0.7	0.2	0.0
0.2	0.4	0.2	0.5	0.4	0.2
71.1	74.4	75.2	73.2	71.5	73.8
0.4	0.7	0.4	0.5	0.5	1.7

fractions of each class have acquired four or more awards for research.

The frequency with which women (Table 14.10) acknowledge one research grant is smaller than that statistic for men, and fluctuates more from year to year. The ratio is persistently in the 3.5 to 5.5% range and does indicate an active interest in research by many members of the female group. About 2.5% of the entire female science graduate sample claim at least two research grant awards.

It appears from these data that attention to research grant opportunities is evident rather early in the career development of science graduates. The situation may be attributable largely to graduate school attendance or employment in nonprofit organizations. More detailed analysis of the data ought to provide enough information for verifying or disproving these hypotheses.

PROFESSIONAL MEMBERSHIPS

Although the impact of membership in professional organizations is difficult to assess, the utility of such membership in publishing articles, writing papers and other activities is likely to be high. The benefits that accrue from membership in these organizations can be characterized as multidimensional insofar as technical expertise, social-professional growth, and economic development are relevant to membership interests.

Science graduates from the liberal arts institutions in this sample furnished data on their status with respect to both regional and national associations. From survey results it appears that the percentage of male graduates belonging to one or more national professional organizations decreases directly with the year of graduation (Table 14.11). There is a year-by-year decrease from 77% of the 1958 class to 34% of the 1967 class, and at the end of the ten-year period the trend is still downward.

The proportion of men belonging to only one such organization remains roughly constant over the period, from 30% of the 1958 class to 26% of the 1967 class, with peaks of 33% and 34% in intervening years. However, a far higher percentage of men graduating from the earliest class (47%) belong to two or more national professional organizations than do those of the 1967 class (8%). The pattern for women graduates is somewhat different (Table 14.12). Year for year there is a smaller percentage of women than men belonging to one or more national organizations; they are within a few percentage points of each other in the latest years, but the differences become greater as the number of years out of college increases. Some 26% of women in the 1967 class belong to one or more national professional organizations, compared with 34% for men, and 44% of the women in the 1958 class do, compared to 77% for men.

Within the ten-year period there is a more striking difference between the percentage of men and women belonging to more than one such organization: while the curve for men graduates continues upward over the whole period, the curve for women peaks from the fifth to the seventh year out of college, decreasing near the end of the period. That is perhaps a logical sequence as women graduates accept the responsibilities of marriage and begin to drop their professional activities.

When membership in only one such organization is considered, there is a much greater similarity between male and female graduates, with both groups of percentages remaining fairly constant over the period. In the class of 1967, one year after graduation, 21% of the women indicated that they belonged to just one professional organization, compared to 26% of the men. In the earliest class (1958), the percentages were 25% of the women and 30% of the men. Throughout the period the two are at most less than ten percentage points apart, and in one year (1965) they are almost identical. Apparently the major factor affecting the overall differences is the tendency of male graduates to join and remain in two or more such organizations.

	1958 N = 870 %	1959 N = 933 %	1960 N = 1014 %	1961 N = 1121 %
Invalid	1.2	0.8	0.5	0.5
No response	5.2	7.1	5.6	6.9
One	29.9	29.6	31.7	32.6
Two	24.6	24.2	22.5	21.3
Three	10.5	10.2	11.0	8.7
Four or more	12.0	10.4	9.1	7.1
Not attempted	16.8	17.8	19.7	23.0
Undertaken, but not achieved	0.0	0.0	0.0	0.0

	1958 N = 365 %	1959 N = 357 %	1960 N = 405 %	1961 N = 468 %
Invalid	0.8	0.0	1.2	0.4
No response	17.8	14.9	15.1	12.6
One	24.7	23.8	22.7	30.3
Two	9.9	10.9	8.6	13.3
Three	4.1	2.8	4.4	3.4
Four or more	4.9	3.9	6.4	2.8
Not attempted	37.8	43.7	41.5	37.2
Undertaken, but not achieved	0.0	0.0	0.0	0.0

TABLE 14.11

Membership of Male Graduates
in National Professional Organizations,
by Year of Graduation

1962 N = 1120 %	1963 N = 1181 %	1964 N = 1285 %	1965 N = 1338 %	1966 N = 1292 %	1967 N = 1363 %
0.5	0.8	0.5	1.0	0.2	0.3
9.1	9.5	10.1	11.3	14.8	16.6
29.8	32.4	33.9	28.9	28.2	25.7
21.0	19.1	14.5	12.9	9.6	7.3
9.2	7.0	5.5	4.1	1.9	0.6
5.9	4.1	3.0	1.1	0.9	0.3
24.6	27.3	32.6	40.7	44.5	49.3
0.0	0.0	0.0	0.0	0.0	0.0

TABLE 14.12

Membership of Female Graduates
in National Professional Organizations,
by Year of Graduation

1962 N = 471 %	1963 N = 543 %	1964 N = 515 %	1965 N = 579 %	1966 N = 562 %	1967 N = 579 %
0.6	0.9	0.6	0.2	0.5	0.2
13.4	12.2	14.8	15.7	17.1	17.3
26.8	27.1	28.2	29.0	26.0	21.4
13.2	14.4	9.3	9.3	8.9	3.6
4.0	3.5	3.3	2.1	1.3	0.5
3.4	2.4	2.1	0.7	0.0	0.2
38.6	39.6	41.8	43.0	46.3	56.8
0.0	0.0	0.0	0.0	0.0	0.0

The substantial interest in professional organizations by women should not be unexpected, considering the data provided in Chapter 13. Although the attrition rate from the labor force is most evident two to four years after graduation, the intervening period is ample time for expression and development of some professional interests. It appears that at least some of those women who are not actively employed are still enrolled in scientific associations.

With regard to membership in local professional groups, there is a definite tendency for both men and women to belong to more national groups than local groups (Tables 14.13 and 14.14).

HONORARY AWARDS AND OTHER PROFESSIONAL RECOGNITION

Honorary awards in this context include postgraduate honorary degrees, medals, certificates of merit, and so on. In this section,

	1958 N = 870 %	1959 N = 933 %	1960 N = 1014 %	1961 N = 1121 %
Invalid	0.9	0.3	0.4	0.5
No response	12.6	16.1	14.5	15.8
One	28.4	28.3	29.1	26.8
Two	14.5	13.2	12.4	10.4
Three	4.8	4.5	3.3	2.7
Four or more	4.6	4.8	2.8	2.7
Not attempted	34.1	32.8	37.6	41.2
Undertaken, but not achieved	0.0	0.0	0.0	0.0

other professional recognition is restricted to regional or professional *Who's Who* membership (Tables 14.15, 14.16) and postdoctoral awards.

Membership in one or more regional or professional *Who's Who* is acknowledged by 8 to 12% of the men from classes of 1958–1960. The number of respondents in the 1967 class who responded similarly is near zero. Six to eight years following graduation appears to be an active period for these individuals. Lower proportions are apparent in the women's sample. The percentages from early classes with memberships are limited to a range of 2.8 to 3.9% for 1958–1960 grads. The proportions from other classes are near zero.

The science graduates have accumulated an impressive number of professional acknowledgements. The 1958–1960 graduating classes contain the largest ratios of men who have been recipients of one or more honorary awards (Table 14.17). The ratio

TABLE 14.13

Membership of Male Graduates
in Local Professional Organizations,
by Year of Graduation

1962 N = 1120 %	1963 N = 1181 %	1964 N = 1285 %	1965 N = 1338 %	1966 N = 1292 %	1967 N = 1363 %
0.3	0.3	0.2	0.5	0.2	0.2
19.5	18.5	18.4	20.0	21.3	21.4
23.7	23.6	19.9	16.1	15.0	13.1
9.0	7.7	7.1	4.9	2.9	1.8
2.1	1.6	1.1	1.6	0.9	0.4
2.0	1.8	1.0	0.5	0.0	0.2
43.6	46.5	52.2	56.5	59.8	63.0
0.0	0.0	0.0	0.0	0.0	0.0

	1958 N = 365 %	1959 N = 357 %	1960 N = 405 %	1961 N = 468 %
Invalid	1.6	0.6	1.2	0.6
No response	19.5	20.7	20.3	16.7
One	20.6	14.9	13.6	22.0
Two	7.1	7.0	6.9	5.3
Three	2.7	2.0	2.7	1.7
Four or more	2.5	3.6	3.2	2.8
Not attempted	46.0	51.3	52.1	50.9
Undertaken, but not achieved	0.0	0.0	0.0	0.0

	1958 N = 870 %	1959 N = 933 %	1060 N = 1014 %	1961 N = 1121 %
Invalid	0.0	0.0	0.0	0.0
No response	88.2	90.4	92.1	95.1
One	9.9	6.7	5.9	3.8
Two	1.4	2.5	1.6	0.8
Three	0.5	0.3	0.3	0.1
Four or more	0.1	0.2	0.1	0.3
Not attempted	0.0	0.0	0.0	0.0
Undertaken, but not achieved	0.0	0.0	0.0	0.0

TABLE 14.14

			Membership of Female Graduates in Local Professional Organizations, by Year of Graduation		

1962 N = 471 %	1963 N = 543 %	1964 N = 515 %	1965 N = 579 %	1966 N = 562 %	1967 N = 579 %
0.4	0.6	1.0	0.4	0.5	0.2
19.3	17.7	17.9	20.0	22.2	19.3
17.6	18.1	17.3	17.4	17.3	13.5
8.7	9.0	5.8	5.7	4.8	2.9
1.9	1.8	3.1	1.7	0.5	0.7
1.9	2.0	2.1	0.9	0.2	0.2
50.1	50.8	52.8	53.9	54.5	63.2
0.0	0.0	0.0	0.0	0.0	0.0

TABLE 14.15

			Membership of Male Graduates in Regional or Professional *Who's Who*, by Year of Graduation		

1962 N = 1120 %	1963 N = 1181 %	1964 N = 1285 %	1965 N = 1338 %	1966 N = 1292 %	1967 N = 1363 %
0.0	0.0	0.0	0.0	0.0	0.0
96.5	98.1	98.8	99.2	99.7	99.6
2.9	1.3	1.0	0.6	0.3	0.4
0.3	0.3	0.1	0.2	0.0	0.1
0.2	0.0	0.0	0.0	0.0	0.0
0.2	0.3	0.2	0.1	0.0	0.0
0.0	0.0	0.0	0.0	0.0	0.0
0.0	0.0	0.0	0.0	0.0	0.0

	1958 N = 365 %	1959 N = 357 %	1960 N = 405 %	1961 N = 468 %
Invalid	0.0	0.0	0.0	0.0
No response	96.2	97.2	96.5	97.0
One	2.5	2.5	3.0	2.4
Two	0.8	0.3	0.5	0.4
Three	0.0	0.0	0.0	0.2
Four or more	0.6	0.0	0.0	0.0
Not attempted	0.0	0.0	0.0	0.0
Undertaken, but not achieved	0.0	0.0	0.0	0.0

	1958 N = 870 %	1959 N = 933 %	1960 N = 1014 %	1961 N = 1121 %
Invalid	0.2	0.0	0.1	0.0
No response	86.3	87.5	89.2	88.1
One	8.7	7.4	6.8	7.7
Two	3.5	3.8	2.6	2.2
Three	0.5	0.8	0.9	1.3
Four or more	0.8	0.6	0.5	0.8
Not attempted	0.0	0.0	0.0	0.0
Undertaken, but not achieved	0.0	0.0	0.0	0.0

TABLE 14.16

			Membership of Female Graduates in Regional or Professional *Who's Who*, by Year of Graduation		
1962 *N = 471* %	*1963* *N = 543* %	*1964* *N = 515* %	*1965* *N = 579* %	*1966* *N = 562* %	*1967* *N = 579* %
0.0	0.0	0.0	0.0	0.0	0.0
98.5	98.7	99.8	99.1	99.5	99.7
0.9	1.1	0.2	0.9	0.5	0.4
0.6	0.2	0.0	0.0	0.0	0.0
0.0	0.0	0.0	0.0	0.0	0.0
0.0	0.0	0.0	0.0	0.0	0.0
0.0	0.0	0.0	0.0	0.0	0.0
0.0	0.0	0.0	0.0	0.0	0.0

TABLE 14.17

			Honorary Awards to Male Graduates, by Year of Graduation		
1962 *N = 1120* %	*1963* *N = 1181* %	*1964* *N = 1285* %	*1965* *N = 1338* %	*1966* *N = 1292* %	*1967* *N = 1363* %
0.0	0.1	0.1	0.0	0.0	0.1
89.3	85.8	87.2	92.5	93.7	96.3
7.0	8.3	8.8	5.5	4.7	2.7
2.8	4.2	2.3	1.2	1.2	0.7
0.4	0.6	0.6	0.6	0.3	0.2
0.6	1.1	0.9	0.3	0.2	0.1
0.0	0.0	0.0	0.0	0.0	0.0
0.0	0.0	0.0	0.0	0.0	0.0

	1958 *N* = 365 %	1959 *N* = 357 %	1960 *N* = 405 %	1961 *N* = 468 %
Invalid	0.0	0.3	0.0	0.0
No response	97.5	97.2	96.5	96.2
One	1.4	2.0	3.0	1.9
Two	0.6	0.6	0.5	1.3
Three	0.3	0.0	0.0	0.2
Four or more	0.3	0.0	0.0	0.4
Not attempted	0.0	0.0	0.0	0.0
Undertaken, but not achieved	0.0	0.0	0.0	0.0

decreases from 13.5 to 10.8% for this period. A small number of the youngest grads (3.7%) have achieved with respect to this criterion at least once and the ratio increases progressively to 12% for the older 1961 graduates.

The data for women graduates differ from the pattern for men to the extent that the proportion of women who acknowledge at least one award is highest for the middle period of 1961–1964, and earlier or later classes include smaller numbers (Table 14.18). During the 1961–1964 period from 3.8 to 4.7% of women from their respective classes fell into this multiple award category. For earlier and for later classes the proportion varies from 2.2 to 3.5%. The higher plateau for middle graduating classes may result, in part, from increasing interest by women in achieving such awards. The more depressed rates for the recent years are probably a function of the relative youth of the graduates.

Postdoctoral awards for study writing or research have in-

TABLE 14.18

		Honorary Awards to Female Graduates, by Year of Graduation			
1962 *N = 471* %	*1963* *N = 543* %	*1964* *N = 515* %	*1965* *N = 579* %	*1966* *N = 562* %	*1967* *N = 579* %
0.0	0.2	0.0	0.0	0.0	0.2
95.8	95.2	96.1	97.1	97.9	97.6
3.6	3.3	2.7	1.4	1.8	1.7
0.6	0.4	1.0	1.0	0.4	0.4
0.0	0.6	0.0	0.5	0.0	0.2
0.0	0.4	0.2	0.0	0.0	0.0
0.0	0.0	0.0	0.0	0.0	0.0
0.0	0.0	0.0	0.0	0.0	0.0

creased substantially during the past ten years. The awards appear to be a convenient device for professional development, but they are limited insofar as only Ph.D. recipients of the current sample will have the opportunity to capitalize on them.

From the earliest three graduating classes of men, the proportion of individuals who have received at least one such award is between 10 and 12.5% (Table 14.19). A relatively stable fraction (7%) of each graduating class from 1958 to 1963 has had one postdoctorate form of recognition. More recent classes contain no male recipients.

Women can be expected to have had fewer postdoctoral awards, insofar as completion of Ph.D. requirements generally takes place later than for men who receive doctorates (Table 14.20). The data presented earlier in this paper are reflected also in the current information. From 3 to 3.6% of the first three graduating classes contained women who said that they had received at least one

	1958 N = 870 %	1959 N = 933 %	1960 N = 1014 %	1961 N = 1121 %
Invalid	0.1	0.0	0.3	0.2
No response	32.1	35.3	32.2	31.1
One	9.3	7.2	7.1	7.1
Two	2.6	2.8	2.4	1.3
Three	0.3	0.2	0.5	0.4
Four or more	0.2	0.1	0.1	0.1
Not attempted	54.5	53.3	55.9	59.3
Undertaken, but not achieved	0.8	1.2	1.6	0.5

	1958 N = 365 %	1959 N = 357 %	1960 N = 405 %	1961 N = 468 %
Invalid	0.0	0.0	0.0	0.2
No response	33.7	34.7	32.8	32.1
One	2.7	1.7	2.0	3.4
Two	0.6	0.8	1.0	0.4
Three	0.3	0.3	0.0	0.0
Four or more	0.0	0.6	0.0	0.0
Not attempted	62.5	61.6	64.0	63.7
Undertaken, but not achieved	0.3	0.3	0.3	0.2

TABLE 14.19

			Postdoctoral Awards to Male Graduates, by Year of Graduation		
1962 N = 1120 %	*1963* N = 1181 %	*1964* N = 1285 %	*1965* N = 1338 %	*1966* N = 1292 %	*1967* N = 1363 %
0.1	0.1	0.0	0.0	0.0	0.0
33.9	32.2	33.0	32.8	35.8	33.0
6.3	6.2	2.5	0.2	0.0	0.0
2.7	0.4	0.2	0.0	0.0	0.0
0.0	0.0	0.1	0.0	0.0	0.0
0.0	0.1	0.1	0.0	0.0	0.0
56.2	59.4	62.7	66.4	64.2	67.0
0.9	1.7	1.5	0.6	0.1	0.0

TABLE 14.20

			Postdoctoral Awards to Female Graduates, by Year of Graduation		
1962 N = 471 %	*1963* N = 543 %	*1964* N = 515 %	*1965* N = 579 %	*1966* N = 562 %	*1967* N = 579 %
0.0	0.0	0.0	0.0	0.0	0.0
35.0	31.9	31.7	30.7	33.1	28.5
1.7	1.3	1.0	0.0	0.0	0.2
0.2	0.4	0.0	0.0	0.0	0.0
0.0	0.0	0.0	0.0	0.0	0.0
0.0	0.0	0.0	0.0	0.0	0.0
62.6	65.9	67.2	69.1	66.9	71.3
0.4	0.6	0.2	0.2	0.0	0.0

postdoctoral award. Most of the individuals received a single award and in the later classes (1965–1967), the number of recipients is, of course, near zero.

SUMMARY

From the preceding discussion, it is evident that for liberal arts science graduates the achievement-productivity structure is complex, with much data conditional on time qualifications and the interaction of sex and type of achievement. For both men and women, professional recognition is frequent and varied. Substantial numbers of both men and women publish articles and present papers at professional meetings. A much smaller but still notable fraction write and publish books. These contributions to literature or science are augmented by inventions, in which men express more interest.

Professional achievements are time-bound especially for the case of book writing and inventing. It is during the period of seven to ten years following graduation that such productivity becomes evident. Early high activity in publishing articles is typical, however, and this is probably influenced by college emphasis on publications for students who have graduate school aspirations.

Professional organizations receive support more frequently from men than from women, and a marked sex differential is apparent in the case of multiple memberships. Women appear to prefer other modes of social development than attending two or more of such groups. A plausible hypothesis is that marriage, church work, or civic activities may generate more interest.

Although the data for inventing are rather limited it is evident that women do not frequently participate in this activity. The extent to which this low level of activity is a function of curriculum (e.g., being in biology rather than physics) is not very clear from the current data and may be examined more carefully with further analysis. Considering the productivity of either men or women, inventions seem to be a stepchild vocation of sorts. If inventions are considered as desirable by the professional communities, and important relative to publications, then information, encouragement and support during undergraduate development appears to be a reasonable suggestion.

A notable ratio of science graduates have received research grants. The extent to which these data are influenced by changes in Federal or private grant policy is not known. It is likely that a high correlation exists between graduates who acknowledge that they have received grants and those who have enrolled in graduate school. Further data analysis is necessary to evaluate this proposal.

The fact that substantial numbers of graduates have received honorary awards within ten years following graduation is likely to be surprising for many individuals. Although such information can occasionally be found in alumni bulletins and college newspapers, its generalized dissemination to the public has not been evident. Systematic data on awards to alumni may be a useful means of providing evidence of expertise to potential students on recruitment and to institutional grant applications, and of evaluating one aspect of the particular institution's productivity.

It would appear, also, that consistent attention to all such awards and public dissemination of results would enhance the institutional attractiveness for many prospective and current students, and would benefit faculty and administrators with regard to other groups interested in the liberal arts enterprise.

15

Postgraduate
Extraprofessional Activities

Activities that are not normally related to professional development might account for much of the science graduates' time and
energies. In the current survey graduates were requested to indicate the extent (never, seldom, frequently, constantly) of their

	1958 N = 870 %	1959 N = 933 %	1960 N = 1014 %	1961 N = 1121 %
Men				
Invalid	0.0	0.0	0.0	0.1
No response	4.8	5.3	6.3	4.6
Never	0.0	0.5	0.4	0.4
Seldom	6.4	4.5	5.9	6.4
Frequently	26.6	27.6	29.8	30.2
Constantly	62.2	62.2	57.6	58.3
	N = 365 %	N = 357 %	N = 405 %	N = 468 %
Women				
Invalid	0.0	0.0	0.0	0.0
No response	7.7	6.4	6.4	6.4
Never	0.0	0.0	1.0	1.1
Seldom	6.9	9.0	6.2	12.8
Frequently	25.5	31.1	24.0	29.5
Constantly	60.0	53.5	62.5	50.2

participation in a diversified list of activities not directly related to employment or development of professional expertise. These included the more common (reading newspapers) as well as the more unique (holding political office) activities.

READING HABITS

Nearly 83% of both men and women responded that they read daily newspapers frequently or constantly (Table 15.1). Women do not differ notably from men in this regard, but a consistent and slightly higher proportion of men gave positive responses. The percentages are roughly stable, although there is evidence which indicates that the older graduates attend more to this medium.

TABLE 15.1

Daily Newspapers Read by Graduates

1962 N = 1120 %	1963 N = 1181 %	1964 N = 1285 %	1965 N = 1338 %	1966 N = 1292 %	1967 N = 1363 %
0.0	0.0	0.0	0.0	0.1	0.1
5.5	5.4	5.3	4.6	5.0	5.4
0.3	0.3	0.5	0.6	0.9	0.2
6.9	7.3	9.3	12.0	10.5	11.7
33.4	35.5	33.7	36.6	37.4	37.8
54.0	51.6	51.2	46.3	46.2	44.8
N = 471 %	N = 543 %	N = 515 %	N = 579 %	N = 562 %	N = 579 %
0.0	0.0	0.0	0.0	0.0	0.0
8.1	8.7	5.8	7.1	8.5	5.0
1.1	0.4	0.8	1.2	0.2	0.5
10.2	12.2	9.5	14.3	13.4	14.5
32.3	32.0	33.2	34.9	33.5	37.0
48.4	46.8	50.7	42.5	44.5	43.0

	1958 N = 870 %	1959 N = 933 %	1960 N = 1014 %	1961 N = 1121 %
Men				
Invalid	0.0	0.0	0.1	0.0
No response	4.9	5.3	6.4	4.6
Never	1.5	1.9	1.3	2.1
Seldom	18.2	17.2	19.5	18.8
Frequently	37.1	36.9	35.6	36.4
Constantly	38.3	38.8	37.1	38.1
	N = 365 %	N = 357 %	N = 405 %	N = 468 %
Women				
Invalid	0.0	0.0	0.0	0.0
No response	7.7	6.4	6.7	6.4
Never	2.2	2.8	2.2	4.3
Seldom	21.4	20.5	19.3	23.3
Frequently	37.5	37.8	31.9	32.7
Constantly	31.2	32.5	40.0	33.3

The data on reading news magazines (Table 15.2) are very similar insofar as proportions for men and women do not differ greatly, but about 4% more of the men affirm frequent or constant attention to news magazines. About 74% of the male graduates during 1958–1967 furnished this response and approximately 70% of the women responded similarly.

Literary magazines receive little attention from this group of respondents (Table 15.3). About 70% of the men and from 60 to 65% of the women graduates during 1958–1967 seldom, if ever, read such journals.

Books that are not particularly related to professional expertise are more popular than literary magazines (Table 15.4). Between 42 and 48% of the men, and from 46 to 52% of the women acknowl-

TABLE 15.2

News Magazines Read by Graduates

1962 N = 1120 %	1963 N = 1181 %	1964 N = 1285 %	1965 N = 1338 %	1966 N = 1292 %	1967 N = 1363 %
0.0	0.0	0.1	0.0	0.0	0.1
5.5	5.7	5.5	4.7	5.1	5.5
2.4	1.5	2.0	1.6	1.9	2.1
18.9	19.3	17.0	18.6	18.7	18.2
40.5	38.4	37.5	42.7	40.1	41.2
32.7	35.1	38.0	32.4	34.3	33.0
N = 471 %	N = 543 %	N = 515 %	N = 579 %	N = 562 %	N = 579 %
0.0	0.0	0.0	0.0	0.2	0.0
7.9	8.8	5.8	7.1	8.4	5.5
1.9	1.8	2.3	1.9	3.0	1.6
22.3	22.3	21.8	22.5	18.0	22.1
38.0	33.9	35.9	38.0	40.6	37.8
29.9	33.2	34.2	30.6	29.9	33.0

edged frequent reading of nonprofessional books. These proportions are roughly constant over the ten years considered. A more marked sex differential is apparent if one considers those graduates who said that they read nonprofessional books constantly. From 12 to 18% of the men did so, and the percentage increases across this range for the time period under examination. The proportion of women who constantly read books unrelated to their professions is in the range of 24 to 34%, and fluctuates considerably from year to year. There is no discernible increase in proportion for women of more recent classes.

The use of a library card is indicative of both professional reading and attention to other substantive topics. Table 15.5 indicates that the number of women who seldom or never capitalize on this service increases during 1958–1967 from 30 to 42%. Fewer (7%) older female graduates respond that they never use a library card, and this percentage increases throughout the ten-year period. The younger men, on the other hand, use library cards more frequently than the graduates of earlier classes. The proportion of those who never use cards is 22% for the 1958 and 1959 classes. This fraction drops to 17% for the 1967 graduates.

CIVIC AND POLITICAL INVOLVEMENT

The majority (more than 70%) of science graduates never attend civic organization meetings such as those required by Kiwanis or

	1958 N = 870 %	1959 N = 933 %	1960 N = 1014 %	1961 N = 1121 %
Men				
Invalid	0.0	0.1	0.0	0.2
No response	5.6	5.8	7.2	5.2
Never	28.6	31.9	30.5	30.8
Seldom	43.6	42.1	40.3	43.3
Frequently	18.7	15.9	17.4	17.0
Constantly	3.5	4.2	4.6	3.6
	N = 365 %	N = 357 %	N = 405 %	N = 468 %
Women				
Invalid	0.0	0.0	0.0	0.0
No response	8.5	7.8	6.7	7.1
Never	22.5	19.6	23.0	25.0
Seldom	40.6	43.1	39.3	41.2
Frequently	21.6	23.5	24.0	21.8
Constantly	6.9	5.9	7.2	4.9

Rotary groups (Table 15.6). Participation of graduates (seldom, frequently, or constantly) decreases for both men (22 to 6%) and women (18 to 4%) through 1958–1967. The highest proportion of frequent and constant involvement in civic organizations is evident for women from the earliest graduating class (10%). Men were not far below this ratio with 7% attending civic organization functions. The smallest fraction of graduates who participate frequently or constantly occurs in the most recent years. Fewer than 2% of the men and women from the 1966 and 1967 classes responded that they were so active.

Education groups (Parents and Teachers Associations) attract many of the older respondents in this sample (Table 15.7). Al-

TABLE 15.3

Literary Magazines Read by Graduates

1962 N = 1120 %	1963 N = 1181 %	1964 N = 1285 %	1965 N = 1338 %	1966 N = 1292 %	1967 N = 1363 %
0.2	0.0	0.1	0.0	0.0	0.0
6.3	6.1	6.1	5.3	5.9	6.2
31.3	32.5	33.4	33.7	32.7	34.0
41.8	39.5	38.3	39.5	37.5	36.8
16.8	18.0	18.4	17.3	19.4	19.2
3.8	3.8	3.7	4.2	4.5	3.8
N = 471 %	N = 543 %	N = 515 %	N = 579 %	N = 562 %	N = 579 %
0.0	0.0	0.0	0.0	0.0	0.0
9.6	9.0	7.4	7.8	9.3	7.3
26.1	25.1	28.2	25.9	25.1	27.1
40.1	38.5	41.4	39.6	34.9	38.7
19.1	21.9	16.9	21.9	22.8	21.1
5.1	5.5	6.2	4.8	8.0	5.9

	1958 N = 870 %	1959 N = 933 %	1960 N = 1014 %	1961 N = 1121 %
Men				
Invalid	0.0	0.0	0.1	0.0
No response	5.2	5.7	6.7	5.0
Never	1.6	2.0	1.5	1.4
Seldom	36.2	35.1	30.0	30.2
Frequently	44.7	42.4	47.7	45.8
Constantly	12.3	14.8	14.0	17.6
	N = 365 %	N = 357 %	N = 405 %	N = 468 %
Women				
Invalid	0.0	0.0	0.3	0.2
No response	8.0	6.7	6.4	6.4
Never	0.3	0.6	0.5	0.9
Seldom	15.6	10.1	13.6	11.5
Frequently	49.3	48.5	48.9	49.6
Constantly	26.9	34.2	30.4	31.4

though only 12% of the 1967 male graduates affirm any involvement in such groups, one-third to one-half of the first three graduating classes did so. From 10 to 17% of the earliest five graduating classes indicated frequent or constant activity.

The female graduates generally participate in educational groups more frequently than men. Of the 1967 graduates 20% responded positively to the question of any activity, and more than half of the earliest three classes occupy this category. The largest fraction (36%) of graduates who acknowledged frequent or constant participation is evident for the women graduates of the 1958 group. The proportion in this category of response decreases to 10% for the 1967 graduates.

Local political organizations appear to be more popular for

TABLE 15.4

		Nonprofessional Books Read by Graduates			
1962 N = 1120 %	*1963* N = 1181 %	*1964* N = 1285 %	*1965* N = 1338 %	*1966* N = 1292 %	*1967* N = 1363 %
0.2	0.0	0.2	0.0	0.0	0.0
6.0	5.7	5.8	5.2	5.0	5.7
1.3	2.0	1.6	1.6	2.6	2.1
34.2	32.6	33.3	35.6	31.2	29.1
43.2	45.0	44.1	43.4	46.8	48.6
15.1	14.8	15.0	14.3	14.4	14.5
N = 471 %	*N = 543* %	*N = 515* %	*N = 579* %	*N = 562* %	*N = 579* %
0.0	0.0	0.0	0.0	0.0	0.0
7.9	8.7	6.8	6.9	8.5	5.9
0.2	0.7	0.2	0.5	0.4	0.4
15.9	15.3	14.8	15.4	14.8	16.2
52.0	46.4	46.6	49.9	46.6	49.1
24.0	28.9	31.7	27.3	29.7	28.5

both men and women than other civic groups (Table 15.8). From 30 to 40% of the women from each graduating class responded that they were active to some extent. Larger proportions of women (8–12%) are to be found among those individuals who participate frequently or constantly. Men also acknowledge a fair interest in these organizations but are less likely to be involved than women, as evidenced by the data for each graduation year.

OTHER ORGANIZED ACTIVITIES

Both men and women in this sample display a notable lack of interest in membership in fraternal organizations (Table 15.9). Nearly 90% of both the men and the women seldom or never pur-

	1958 $\tilde{N} = 870$ %	1959 $N = 933$ %	1960 $N = 1014$ %	1961 $N = 1121$ %
Men				
Invalid	0.1	0.0	0.1	0.3
No response	5.6	5.8	7.0	5.1
Never	22.1	22.3	20.4	19.9
. Seldom	32.3	32.8	32.3	33.6
Frequently	29.0	28.0	27.3	28.5
Constantly	10.9	11.2	12.9	12.7
	$N = 365$ %	$N = 357$ %	$N = 405$ %	$N = 468$ %
Women				
Invalid	0.0	0.0	0.0	0.0
No response	8.2	7.0	7.2	6.6
Never	7.1	7.3	9.1	10.9
Seldom	23.0	22.1	27.2	26.7
Frequently	38.4	34.7	34.6	33.3
Constantly	23.3	28.9	22.0	22.4

	1958 $N = 870$ %	1959 $N = 933$ %	1960 $N = 1014$ %	1961 $N = 1121$ %
Men				
Invalid	0.0	0.0	0.0	0.1
No response	6.1	5.8	6.9	5.7
Never	72.3	73.4	75.6	75.7
Seldom	14.5	12.9	11.1	12.0
Frequently	4.8	4.4	4.2	4.3
Constantly	2.3	3.5	2.1	2.2
	$N = 365$ %	$N = 357$ %	$N = 405$ %	$N = 468$ %
Women				
Invalid	0.3	0.6	0.5	0.2
No response	9.3	8.4	7.7	7.7
Never	72.3	75.6	72.8	78.6
Seldom	8.2	8.1	9.9	6.8
Frequently	7.7	5.0	7.4	4.9
Constantly	2.2	2.2	1.7	1.7

TABLE 15.5

Use of Library Cards by Graduates

1962 N = 1120 %	1963 N = 1187 %	1964 N = 1285 %	1965 N = 1338 %	1966 N = 1292 %	1967 N = 1363 %
0.0	0.0	0.0	0.2	0.0	0.1
6.2	6.2	5.7	5.2	5.7	6.2
19.5	22.9	20.1	18.0	18.3	16.8
32.6	31.7	29.7	29.7	29.0	29.4
29.3	27.5	31.4	31.9	31.9	32.7
12.5	11.8	13.2	15.1	15.1	15.0
N = 471 %	N = 543 %	N = 515 %	N = 579 %	N = 562 %	N = 579 %
0.0	0.0	0.2	0.0	0.0	0.2
9.1	9.4	7.4	7.3	9.1	6.4
11.7	12.0	15.0	15.4	14.6	15.7
29.1	34.6	29.9	30.6	27.2	26.3
30.2	29.8	32.0	30.6	32.6	34.9
20.0	14.2	15.5	16.2	16.6	16.6

TABLE 15.6

Civic Organizations Attended by Graduates

1962 N = 1120 %	1963 N = 1181 %	1964 N = 1285 %	1965 N = 1338 %	1966 N = 1292 %	1967 N = 1363 %
0.0	0.0	0.0	0.0	0.0	0.0
6.6	6.2	5.8	5.2	5.8	6.2
78.7	83.3	82.7	86.3	86.0	87.5
9.6	7.0	9.0	6.3	6.6	5.6
3.5	2.7	1.9	1.7	1.5	0.6
1.6	0.8	0.6	0.5	0.2	0.2
N = 471 %	N = 543 %	N = 515 %	N = 579 %	N = 562 %	N = 579 %
0.4	0.0	0.0	0.2	0.2	0.0
9.3	9.8	8.0	8.3	9.8	6.9
81.3	82.0	83.3	86.9	85.6	87.9
4.5	5.5	4.9	4.0.	3.6	3.8
3.8	1.8	2.7	0.7	0.7	1.4
0.6	0.9	1.2	0.0	0.2	0.0

	1958 N = 870 %	1959 N = 933 %	1960 N = 1014 %	1961 N = 1121 %
Men				
Invalid	0.0	0.1	0.0	0.0
No response	5.8	5.9	6.8	5.4
Never	48.9	52.0	58.9	62.8
Seldom	28.2	27.7	24.0	20.9
Frequently	15.1	10.7	8.4	8.7
Constantly	2.2	3.6	2.0	2.2
	N = 365 %	N = 357 %	N = 405 %	N = 468 %
Women				
Invalid	0.0	0.0	0.0	0.0
No response	9.6	9.2	7.9	8.8
Never	29.3	38.1	40.3	55.6
Seldom	24.7	18.5	21.7	16.7
Frequently	29.3	27.2	23.2	15.0
Constantly	7.1	7.0	6.9	4.1

	1958 N = 870 %	1959 N = 933 %	1960 N = 1014 %	1961 N = 1121 %
Men				
Invalid	0.0	0.0	0.0	0.0
No response	5.5	5.9	7.0	5.8
Never	61.2	64.7	62.7	62.5
Seldom	25.9	21.9	24.1	23.4
Frequently	6.1	6.9	5.5	6.8
Constantly	1.4	0.6	0.7	1.5
	N = 365 %	N = 357 %	N = 405 %	N = 468 %
Women				
Invalid	0.0	0.0	0.0	0.0
No response	10.1	8.4	7.2	7.3
Never	49.6	52.4	58.0	57.3
Seldom	28.2	29.7	25.9	24.8
Frequently	10.1	8.7	8.2	9.6
Constantly	1.9	0.8	0.7	1.1

TABLE 15.7

Participation of Graduates in Educational Groups

1962 N = 1120 %	1963 N = 1181 %	1964 N = 1285 %	1965 N = 1338 %	1966 N = 1292 %	1967 N = 1363 %
0.2	0.2	0.0	0.0	0.0	0.1
6.3	6.4	6.1	5.4	6.0	6.2
68.7	76.0	77.8	78.7	80.0	81.8
14.6	11.8	10.7	9.5	9.5	7.0
8.0	4.2	3.8	4.7	3.2	3.6
2.2	1.4	1.6	1.7	1.3	1.3
N = 471 %	N = 543 %	N = 515 %	N = 579 %	N = 562 %	N = 579 %
0.2	0.0	0.0	0.0	0.0	0.0
10.0	9.8	7.8	8.6	10.0	6.6
59.5	59.1	62.9	66.2	67.6	73.5
16.4	16.0	13.4	12.4	9.3	9.5
11.0	12.2	11.8	10.5	9.6	8.1
3.0	3.0	4.1	2.3	3.6	2.3

TABLE 15.8

Participation of Graduates in Local Political Groups

1962 N = 1120 %	1963 N = 1181 %	1964 N = 1285 %	1965 N = 1338 %	1966 N = 1292 %	1967 N = 1363 %
0.0	0.0	0.0	0.0	0.0	0.0
6.6	5.9	6.2	5.3	5.9	6.9
65.8	69.0	65.9	67.0	67.7	64.7
19.8	18.1	20.5	20.6	20.5	20.7
6.6	6.2	6.6	6.1	5.0	6.8
1.2	0.8	0.7	1.1	0.9	1.0
N = 471 %	N = 543 %	N = 515 %	N = 579 %	N = 562 %	N = 579 %
0.0	0.0	0.2	0.0	0.0	0.0
9.6	9.4	7.4	8.5	10.0	7.3
58.4	59.9	62.9	61.7	59.4	61.5
22.3	23.0	23.1	22.8	22.1	23.1
7.9	7.0	5.6	6.7	7.8	7.6
1.9	0.7	0.8	0.4	0.7	0.5

	1958 N = 870 %	1959 N = 933 %	1960 N = 1014 %	1961 N = 1121 %
Men				
Invalid	0.1	0.0	0.0	0.0
No response	6.0	6.1	6.9	6.1
Never	79.2	80.7	81.4	82.4
Seldom	10.8	10.3	9.0	9.1
Frequently	2.4	1.9	2.4	1.5
Constantly	1.5	1.0	0.4	0.9
	N = 365 %	N = 357 %	N = 405 %	N = 468 %
Women				
Invalid	0.0	0.0	0.3	0.2
No response	11.2	8.4	7.4	7.5
Never	81.6	83.2	84.7	85.7
Seldom	4.7	5.9	4.7	3.4
Frequently	2.2	2.2	2.2	2.4
Constantly	0.3	0.3	0.7	0.9

sue this activity. Within each of these groups, the "never" category is occupied by most of the respondents each year.

Nonfraternal service groups (Red Cross, etc.) receive only a bit more interest (Table 15.10). From 84 to 88% of the men seldom, if ever, participate in related activities. Women do become involved more frequently, but the majority (52–74%) are quite inactive. The proportion of women in the most recent graduating class who seldom or never participate in service groups is large (86%) by comparison to the earliest group (71%). As age increases, it appears that women are somewhat more likely to engage in service group activities.

TABLE 15.9

					Fraternal Groups Attended by Graduates
1962	*1963*	*1964*	*1965*	*1966*	*1967*
N = 1120	*N = 1181*	*N = 1285*	*N = 1338*	*N = 1292*	*N = 1363*
%	%	%	%	%	%
0.0	0.0	0.0	0.0	0.0	0.0
6.6	6.4	6.1	5.2	6.0	6.2
80.7	82.1	81.3	82.3	81.7	80.0
9.6	8.6	9.2	8.5	8.8	8.4
2.2	2.2	2.7	3.2	2.8	4.4
0.9	0.7	0.7	0.7	0.7	0.9
N = 471	*N = 543*	*N = 515*	*N = 579*	*N = 562*	*N = 579*
%	%	%	%	%	%
0.0	0.0	0.0	0.0	0.0	0.2
9.6	9.8	8.0	8.5	9.8	7.3
84.9	82.5	85.8	87.6	85.8	87.1
4.3	5.2	4.3	2.4	3.7	3.6
1.1	2.4	1.8	1.4	0.7	1.7
0.2	0.2	0.2	0.2	0.0	0.2

The majority of men never serve in church or synagogue activities (Table 15.11). The lowest fraction of any participation occurs for the more recent graduating classes in which nearly 65% of the respondents indicate no service of this type. Less than half of the earliest graduates acknowledge seldom, frequent or constant activity. As one may expect, the data for women are not quite so extreme. Although recent graduates are fairly inactive (58% say that they never serve in churches or synagogues), the proportions of earlier classes who frequently or constantly participate is in the 30 to 40% range. The decrement in this latter category is marked for the 1958–1967 period. From 40 to 17% of the 1958 and 1967

	1958 N = 870 %	1959 N =933 %	1960 N = 1014 %	1961 N = 1121 %
Men				
Invalid	0.1	0.2	0.2	0.3
No response	6.2	6.3	6.9	6.2
Never	66.2	67.3	70.0	69.9
Seldom	18.5	17.9	15.7	16.2
Frequently	6.7	6.0	5.5	6.1
Constantly	2.3	2.3	1.7	1.4
	N = 365 %	N = 357 %	N = 405 %	N = 468 %
Women				
Invalid	0.0	0.0	0.5	0.0
No response	9.6	8.4	7.9	8.1
Never	51.8	54.6	55.3	58.1
Seldom	19.2	21.0	19.5	19.4
Frequently	14.5	13.5	12.6	11.1
Constantly	4.9	2.5	4.2	3.2

	1958 N = 870 %	1959 N = 933 %	1960 N = 1014 %	1961 N = 1121 %
Men				
Invalid	0.0	0.1	0.0	0.1
No response	5.5	6.3	6.5	5.2
Never	50.1	49.5	51.7	55.3
Seldom	20.6	19.8	20.3	21.1
Frequently	13.1	13.8	11.6	11.1
Constantly	10.7	10.4	9.9	7.2
	N = 365 %	N = 357 %	N = 405 %	N = 468 %
Women				
Invalid	0.0	0.0	0.3	0.0
No response	8.2	8.1	7.2	6.8
Never	35.6	43.4	44.9	53.6
Seldom	17.0	18.5	18.3	16.2
Frequently	20.0	16.0	16.8	14.1
Constantly	19.2	14.0	12.6	9.2

TABLE 15.10

Participation of Graduates in Service Groups

1962 N = 1120 %	1963 N = 1181 %	1964 N = 1285 %	1965 N = 1338 %	1966 N = 1292 %	1967 N = 1363 %
0.5	0.4	0.0	0.2	0.2	0.1
7.1	6.4	6.5	5.8	6.0	6.4
74.7	76.5	78.4	77.6	78.8	79.8
13.0	11.6	10.4	12.0	10.7	9.6
3.8	3.6	4.2	3.6	3.5	3.6
1.0	1.5	0.4	0.9	0.9	0.5

N = 471 %	N = 543 %	N = 515 %	N = 579 %	N = 562 %	N = 579 %
0.2	0.2	0.0	0.0	0.0	0.0
9.8	10.1	8.5	9.0	10.5	7.3
62.0	63.2	63.5	72.4	69.0	73.8
17.4	17.1	16.9	12.3	13.9	12.1
8.5	7.6	8.9	5.2	6.1	5.2
2.1	1.8	2.1	1.2	0.5	1.7

TABLE 15.11

Service of Graduates in Churches or Synagogues

1962 N = 1120 %	1963 N = 1181 %	1964 N = 1285 %	1965 N = 1338 %	1966 N = 1292 %	1967 N = 1363 %
0.0	0.0	0.1	0.0	0.0	0.0
6.2	5.9	5.8	5.2	5.8	6.2
63.8	62.0	64.3	66.9	67.7	64.3
16.0	18.0	17.6	17.4	17.8	21.3
9.1	8.7	7.1	6.4	6.2	5.4
5.0	5.3	5.1	4.2	2.5	2.8

N = 471 %	N = 543 %	N = 515 %	N = 579 %	N = 562 %	N = 579 %
0.0	0.0	0.0	0.0	0.0	0.2
8.5	9.6	7.6	8.3	9.8	7.8
49.7	54.0	56.3	61.8	59.6	57.9
19.5	16.8	19.0	16.9	16.4	17.1
13.6	10.1	9.3	7.8	8.7	11.7
8.7	9.6	7.8	5.2	5.5	5.4

women graduates, respectively, affirmed a typically inactive interest in church-related work.

Club membership activities attract a substantial proportion of the science graduates (Table 15.12). More than half the graduates acknowledge some participation or more frequent involvement in club activities. Between 25 and 30% of the men do so frequently or constantly. The proportions represent the limiting values with respect to the time period considered, and there appears to be a very slight trend in the direction of less frequent activity for the more recent graduates. The frequent or constant participation by women in club activities decreases from 33 to 22% during 1958–1967. For both men and women, the category of "seldom par-

	1958 N = 870 %	1959 N = 933 %	1960 N = 1014 %	1961 N = 1121 %
Men				
Invalid	0.0	0.0	0.2	0.4
No response	6.0	6.4	7.8	5.8
Never	40.9	43.4	42.1	41.5
Seldom	23.2	20.4	23.0	21.1
Frequently	19.7	21.1	19.6	20.7
Constantly	10.2	8.7	7.3	10.6
	N = 365 %	N = 357 %	N = 405 %	N = 468 %
Women				
Invalid	0.0	0.0	0.0	0.2
No response	9.0	7.8	7.2	8.1
Never	35.3	39.2	37.5	38.9
Seldom	22.2	22.1	20.7	21.6
Frequently	25.2	21.0	21.7	23.1
Constantly	8.2	9.8	12.8	8.1

ticipate" comprises about one-fifth of the graduates in each class.

Some science and mathematics graduates have played leadership roles in civic, educational or service groups (Table 15.13). Of all those who graduated in the classes of 1958 and 1959, about 22% of the men and 30% of the women have held office in such organizations. The proportion of more recent graduates who indicated that they seldom, frequently or constantly held office is smaller. Only about 6% of women graduates of the 1966 classes, and 10% of the men had official leadership roles.

A typically small percentage of male and female graduates indicated that they had held a local political office (Table 15.14).

TABLE 15.12

Participation of Graduates as Club Members

1962 N = 1120 %	1963 N = 1181 %	1964 N = 1285 %	1965 N = 1338 %	1966 N = 1292 %	1967 N = 1363 %
0.2	0.1	0.2	0.0	0.1	0.0
6.5	6.6	6.6	6.1	6.0	6.6
43.8	46.2	46.2	48.0	47.7	46.7
21.7	18.5	19.1	17.3	18.9	18.8
19.7	20.3	20.6	22.1	20.3	19.2
8.1	8.3	7.4	6.7	7.0	8.7
N = 471 %	N = 543 %	N = 515 %	N = 579 %	N = 562 %	N = 579 %
0.0	0.0	0.0	0.0	0.0	0.5
10.0	10.1	8.5	8.3	10.1	8.1
41.2	42.7	45.4	48.5	50.2	53.0
17.6	21.2	17.1	18.3	16.9	16.1
23.6	17.3	19.8	16.6	15.8	15.7
7.6	8.7	9.1	8.3	6.9	6.6

	1958 N = 870 %	1959 N = 933 %	1960 N = 1014 %	1961 N = 1121 %
Men				
Invalid	0.0	0.1	0.2	0.1
No response	6.2	6.3	6.7	5.9
Never	69.5	70.1	75.4	74.9
Seldom	12.9	12.9	9.2	9.9
Frequently	8.2	7.5	5.9	7.4
Constantly	3.2	3.1	2.6	1.8
	N = 365 %	N = 357 %	N = 405 %	N = 468 %
Women				
Invalid	0.0	0.0	0.3	0.0
No response	9.0	9.0	7.7	8.1
Never	54.3	62.2	62.5	70.5
Seldom	17.0	12.0	13.3	9.4
Frequently	16.4	12.0	12.8	9.2
Constantly	3.3	4.8	3.5	2.8

	1958 N = 870 %	1959 N = 933 %	1960 N = 1014 %	1961 N = 1121 %
Men				
Invalid	0.0	0.1	0.1	0.1
No response	5.9	5.8	7.0	5.8
Never	90.7	92.1	90.5	92.1
Seldom	1.8	1.3	1.7	1.6
Frequently	0.9	0.2	0.5	0.4
Constantly	0.7	0.5	0.2	0.1
	N = 365 %	N = 357 %	N = 405 %	N = 468 %
Women				
Invalid	0.3	0.0	0.3	0.0
No response	9.3	7.8	7.7	7.7
Never	88.8	91.9	90.4	91.5
Seldom	1.4	0.3	1.2	0.2
Frequently	0.0	0.0	0.3	0.6
Constantly	0.3	0.0	0.3	0.0

TABLE 15.13

Offices Held by Graduates in Civic, Fraternal, or Educational Groups

1962 N = 1120 %	1963 N = 1181 %	1964 N = 1285 %	1965 N = 1338 %	1966 N = 1292 %	1967 N = 1363 %
0.1	0.0	0.0	0.1	0.1	0.0
6.5	6.4	6.1	5.2	6.2	6.6
78.0	81.3	81.8	82.1	82.7	83.6
8.3	7.4	7.1	7.0	6.7	6.2
5.4	3.6	4.1	4.3	3.7	2.6
1.7	1.4	1.0	1.4	0.6	1.0
N = 471 %	N = 543 %	N = 515 %	N = 579 %	N = 562 %	N = 579 %
0.0	0.0	0.0	0.0	0.0	0.2
9.8	10.1	7.8	8.6	10.1	7.4
75.2	75.7	80.4	82.4	83.8	87.6
8.1	7.7	6.8	4.3	3.9	2.6
5.9	5.3	3.7	4.2	1.4	2.3
1.1	1.1	1.4	0.5	0.7	0.0

TABLE 15.14

Local Political Offices Held by Graduates

1962 N = 1120 %	1963 N = 1181 %	1964 N = 1285 %	1965 N = 1338 %	1966 N = 1292 %	1967 N = 1363 %
0.2	0.0	0.1	0.1	0.1	0.2
6.3	6.1	5.9	5.7	6.1	6.6
92.1	92.3	93.0	92.8	92.5	92.2
1.2	1.1	0.9	0.8	1.2	0.8
0.3	0.1	0.2	0.5	0.1	0.2
0.0	0.4	0.0	0.2	0.0	0.0
N = 471 %	N = 543 %	N = 515 %	N = 579 %	N = 562 %	N = 579 %
0.0	0.2	0.0	0.0	0.0	0.0
10.2	9.6	8.0	8.3	9.8	7.4
88.8	88.6	91.3	91.4	89.0	91.9
0.4	1.3	0.8	0.4	1.3	0.5
0.4	0.2	0.0	0.0	0.0	0.2
0.2	0.2	0.0	0.0	0.0	0.0

	1958 N = 870 %	1959 N = 933 %	1960 N = 1014 %	1961 N = 1121 %
Men				
Invalid	0.1	0.0	0.4	0.2
No response	5.8	6.7	7.1	6.1
Never	62.1	67.0	68.8	71.5
Seldom	21.0	18.3	16.6	16.3
Frequently	9.0	6.8	5.4	5.1
Constantly	2.1	1.3	1.7	0.9
	N = 365 %	N = 357 %	N = 405 %	N = 468 %
Women				
Invalid	0.0	0.3	0.0	0.0
No response	9.0	8.4	7.9	7.5
Never	56.4	59.7	61.2	67.3
Seldom	22.2	20.7	18.8	18.4
Frequently	9.9	9.0	10.9	4.9
Constantly	2.5	2.0	1.2	1.9

	1958 N = 870 %	1959 N = 933 %	1960 N = 1014 %	1961 N = 1121 %
Men				
Invalid	0.0	0.0	0.1	0.0
No response	5.3	5.7	6.6	4.8
Never	3.1	3.3	4.3	3.2
Seldom	39.1	37.4	37.3	39.4
Frequently	46.8	47.2	45.7	46.4
Constantly	5.8	6.4	6.0	6.2
	N = 365 %	N = 357 %	N = 405 %	N = 468 %
Women				
Invalid	0.0	0.0	0.0	0.0
No response	8.2	7.3	6.4	6.6
Never	0.3	1.7	1.5	2.1
Seldom	35.6	25.2	28.9	32.3
Frequently	46.6	54.1	52.1	47.4
Constantly	9.3	11.8	11.1	11.5

TABLE 15.15

Appointment of Graduates to Community Committees

1962 N = 1120 %	1963 N = 1181 %	1964 N = 1285 %	1965 N = 1338 %	1966 N = 1292 %	1967 N = 1363 %
0.1	0.1	0.0	0.1	0.2	0.2
6.5	6.2	6.0	5.2	5.9	6.5
74.6	81.4	81.7	82.0	83.7	84.7
14.2	8.4	8.9	10.0	8.2	6.6
3.9	2.9	3.0	2.5	1.6	1.7
0.6	1.1	0.5	0.2	0.4	0.4
N = 471 %	N = 543 %	N = 515 %	N = 579 %	N = 562 %	N = 579 %
0.4	0.0	0.2	0.0	0.2	0.2
9.6	9.8	7.8	8.6	10.0	7.4
71.1	74.4	76.9	78.1	77.6	83.6
13.8	11.4	11.7	10.7	10.1	7.3
4.5	3.7	2.5	2.3	2.0	1.0
0.6	0.7	1.0	0.4	0.2	0.5

TABLE 15.16

Cultural Events Attended by Graduates

1962 N = 1120 %	1963 N = 1181 %	1964 N = 1285 %	1965 N = 1338 %	1966 N = 1292 %	1967 N = 1363 %
0.0	0.1	0.0	0.0	0.0	0.0
5.5	5.7	5.8	4.8	5.3	5.9
5.1	5.3	4.9	6.7	7.5	7.0
39.5	42.0	38.4	40.1	41.1	41.8
43.5	41.5	44.4	43.1	40.2	39.9
6.4	5.5	6.6	5.5	5.9	5.3
N = 471 %	N = 543 %	N = 515 %	N = 579 %	N = 562 %	N = 579 %
0.0	0.0	0.0	0.0	0.0	0.0
8.1	9.0	6.0	7.3	8.7	5.7
2.1	1.7	3.1	3.1	2.7	2.9
32.3	31.5	33.2	34.7	28.1	29.4
48.4	47.2	46.2	46.3	49.5	52.0
9.1	10.7	11.5	8.6	11.0	10.0

Less than 4% of the earliest male graduates, and less than 2% of the earliest female graduates responded positively to this question. Moreover, these percentages decrease to near zero in the later classes.

Community service on committees attracts a larger percentage of graduates (Table 15.15). The proportion of men and women involved in this activity on a seldom or frequent basis approaches one-third of the earliest class. Women are more likely to be associated with this aspect of postgraduate activity.

Cultural events (opera, dance, etc.) are frequently attended by approximately the same percentages of male and female graduates, although somewhat fewer men attend these events constantly (Table 15.16). The percentage of each graduating class who responded positively remains fairly constant across the ten years considered. In more recent years, women have shown more frequent attendance than men.

SUMMARY

Reading habits of the science and mathematics graduates reflect a mixed pattern of behavior. Newspapers and periodical news magazines are read frequently or constantly by a large majority of respondents regardless of the particular graduating class. A minority of graduates read literary magazines but the proportion of those who do so (20%) is fairly stable across classes. Nonprofessional books are read frequently or constantly by a large majority of graduates, and the ratio is generally higher for women than for men.

Civic organizations, such as Kiwanis and Rotary clubs, are not generally attractive to science graduates. Frequent participation in political activities is acknowledged by a minority, however, and the more infrequent activity of the other graduates suggests that about half of the graduates attend political activities at least occasionally. Education groups, such as the P.T.A., attract wider attention from the science graduates, particularly the women, with over half of the earliest classes indicating some participation. A very small minority participate with any frequency in fraternal organizations. Playing a service-oriented role in churches, synagogues or service groups attracts a larger percentage of the graduates. Club membership appears to be most popular insofar as nearly half the graduates acknowledge some participation.

For each of these activities, the proportion of participating respondents decreases as one considers later graduating classes. Sex differentials are most evident in the category of church work (higher proportions of women), and club member activities (larger percentages of men are involved).

Leadership roles, in the sense of holding office in civic, educational or fraternal organizations have been fulfilled by about 25% of the earliest graduates. This statistic includes respondents who seldom do so as well as the more active groups. More recent graduates are less active. Fewer than 4% of the men and 2% of the women indicated that they have held local political office; the small percentage is fairly stable and decreases as one considers the later graduates.

16

Graduate Opinions of
Undergraduate Education

The science graduates provided information on some of their own perceptions of the quality and utility of their undergraduate training. Rather general student concerns, such as competitiveness of graduate schools, as well as specific topics, such as strengths and weaknesses of undergraduate training, were evaluated by the respondents. The data are presented here in the context of major discipline categories, type of college, and year of graduation. There are three important constraints in evaluating these data. The graduate schools about which students furnish perceptions differ in quality and in substantive emphasis, and presumably these factors will influence the graduates' opinions. The quality of the graduates themselves varies, although they are likely to be among the achievers in their respective classes, and this factor can also be expected to influence judgments. Finally, differences in perceptions which are classified by major discipline categories are attributable to sex differences as men and women enter the different disciplines at different rates.

With regard to specific strengths and weaknesses of their undergraduate programs, the graduates provided opinions on the adequacy of preparation in laboratory technique, advanced coursework, and research orientation. More general aspects of faculty quality, stimulation of undergraduate creativity, and development of student interest were also evaluated.

Considering the opinions about preparation in laboratory technique, almost one-half of the chemistry and pre-med graduates said that their preparation was strong and about 40% stated that this preparation was adequate (Table 16.1a). Among biology graduates, approximately equal percentages (40%) described the preparation in laboratory technique as strong or adequate. Math and physics majors were least likely to describe their preparation as strong and most likely to state that it was weak. The 30% of the math graduates who furnished no opinion in this category probably had limited contact with laboratory-oriented courses. Approximately one-third of all graduates encouraged increased emphasis on undergraduate laboratory techniques.

Graduates' opinions about preparation for advanced courses indicates a modal response of adequate for each discipline (Table 16.1a). The opinion that such preparation was weak was shared by a smaller percentage of graduates than those who considered it to be strong. Physics majors were most likely to describe their preparation in advanced coursework as weak, and recommend increased preparation in these courses. Almost half of the graduates said that preparation in advanced courses should be strengthened.

When requested to provide their perceptions on research orientation of their undergraduate training (Table 16.1b), the modal response was that such orientation was adequate or weak. More than 22% of biology majors and 27% of chemistry graduates maintained that research orientation was strong, with smaller percentages of physics, pre-med or math students providing such endorsement. Chemistry majors (28%) were least likely and pre-med graduates most likely (42%) to describe research orientation in their undergraduate programs as weak. From 40 to 50% of all graduates recommended increased emphasis on a research viewpoint in undergraduate training.

Graduates affirmed high faculty quality more frequently than any other feature of their graduate training (Table 16.1b). More than 50% of the graduates in each major discipline category af-

Preparation in laboratory technique
No response
Strong
Adequate
Weak
No opinion
Invalid
Should be increased
Should be decreased
Preparation in advanced courses
No response
Strong
Adequate
Weak
No opinion
Invalid
Should be increased
Should be decreased

firmed the strength of the faculty; from 88 to 92% felt that quality was either adequate or high. Approximately one-third of the respondents said that increases in quality were unwarranted.

Consider now the more ambiguous areas with which graduates and administrative personnel might be concerned. Has the graduate experience provided students with a diversified knowledge? As far as graduates can judge, has student creativity and development of interest been emphasized to a great enough extent in undergraduate training?

TABLE 16.1a

Graduates' Opinions on Adequacy of Types of Training
in Undergraduate Preparation,
by Major Subject

Biology N = 5996 %	Chemistry N = 3362 %	Math N = 3666 %	Physics N = 2045 %	Pre-med N = 1329 %
6.7	5.6	11.1	5.8	5.9
39.9	48.3	29.8	30.1	46.1
42.6	39.7	35.2	47.9	40.9
9.9	5.6	12.9	13.9	5.9
0.8	0.7	10.9	2.2	1.1
0.1	0.0	0.0	0.2	0.0
41.2	33.8	27.3	40.1	28.1
2.4	2.4	1.0	2.1	6.3
7.2	5.9	6.0	6.0	6.6
34.0	31.2	29.6	24.4	36.7
40.6	43.1	43.5	42.7	39.1
14.1	17.0	17.3	23.9	13.2
4.0	2.7	3.6	2.8	4.3
0.1	0.1	0.0	0.2	0.1
48.3	47.4	45.5	51.6	44.4
0.5	0.4	0.6	0.7	1.0

Table 16.1c shows that 36–50% of the graduates felt that the
diversity of training was strong. Smaller percentages of math and
physics majors agreed with the perspective that training provided
strong "breadth of knowledge" than did members of the other
groups. A substantial ratio (39–47%) maintained that the emphasis
on this subject was adequate, but approximately one-third suggested increased attention to a more well-rounded education.

Undergraduate emphasis on student creativity is generally said
to be adequate by graduates in all disciplines considered, but

about one-fifth of the students claimed that their training was weak in this regard and an additional 22–28% furnished the opinion that emphasis was strong. More than 40% of the graduates encouraged increased strengthening of the policy to cultivate student creativity.

Weaknesses associated with the development of student interest were acknowledged less frequently than with student creativity. Again, the modal response is that the emphasis is adequate, but should be given increased attention. Mathematics, physics and pre-med majors were less likely to say that this is a strong feature

Research orientation
  No response
  Strong
  Adequate
  Weak
  No opinion
  Invalid
  Should be increased
  Should be decreased
Quality of faculty
  No response
  Strong
  Adequate
  Weak
  No opinion
  Invalid
  Should be increased
  Should be decreased

of undergraduate training, and more likely to maintain that it is weak.

Only a few major differences among opinions about the above issues are evident when comparisons across college type categories (Tables 16.2a,b,c) are made. Graduates of men's colleges appear somewhat more likely to judge laboratory preparation, advanced course preparation, faculty quality, and breadth of knowledge to be strong. The range of percentages is typically 5–10% over analogous response rates in the other categories. They are also less likely to judge these issues as weak by a range of 2–3%

TABLE 16.1b

Graduates' Opinions on Research Orientation and Faculty Quality, by Major Subject

Biology N = 5996 %	Chemistry N = 3362 %	Math N = 3666 %	Physics N = 2045 %	Pre-med N = 1329 %
7.0	5.9	8.0	6.2	6.4
22.2	27.4	10.8	17.8	11.4
35.2	35.5	29.7	34.9	33.9
32.2	27.8	34.8	35.8	42.2
3.5	3.4	16.6	5.4	6.1
0.0	0.1	0.1	0.0	0.0
48.8	44.7	40.5	46.0	44.5
1.5	1.5	1.3	1.8	3.1
6.7	5.6	5.6	5.5	5.9
58.9	63.0	52.4	52.2	65.4
30.0	28.8	37.0	35.4	26.0
3.7	2.1	4.4	5.8	2.3
0.4	0.4	0.5	0.8	0.2
0.3	0.1	0.2	0.2	0.2
36.0	29.1	32.7	34.6	34.9
0.1	0.0	0.1	0.1	0.1

Breadth of knowledge
  No response
  Strong
  Adequate
  Weak
  No opinion
  Invalid
  Should be increased
  Should be decreased
Development of student creativity
  No response
  Strong
  Adequate
  Weak
  No opinion
  Invalid
  Should be increased
  Should be decreased
Development of student interest
  No response
  Strong
  Adequate
  Weak
  No opinion
  Invalid
  Should be increased
  Should be decreased

TABLE 16.1c

Graduates' Opinions on Breadth of Knowledge and Development
of Student Interest and Creativity,
by Major Subject

Biology N = 5996 %	Chemistry N = 3362 %	Math N = 3666 %	Physics N = 2045 %	Pre-med N = 1329 %
7.0	6.0	6.0	5.9	6.0
47.9	41.5	37.0	36.1	49.9
39.4	45.2	46.6	46.1	38.6
4.8	6.2	8.7	10.4	4.9
0.9	1.0	1.6	1.4	0.6
0.0	0.1	0.1	0.1	0.0
34.7	32.2	33.3	36.0	33.7
0.6	0.8	0.5	1.1	0.6
7.0	6.0	6.4	6.1	6.4
27.6	26.7	21.8	26.1	22.0
42.4	44.2	44.2	42.1	47.9
18.9	17.9	21.1	19.9	20.5
4.0	5.2	6.5	5.9	3.2
0.1	0.0	0.1	0.1	0.1
45.9	41.6	40.6	40.8	45.6
0.2	0.2	0.2	0.4	0.5
7.0	6.0	6.1	6.0	6.3
40.5	39.7	27.9	34.0	33.9
38.2	40.1	46.2	40.6	43.9
12.3	11.7	16.9	16.1	14.1
2.0	2.5	3.0	3.2	1.9
0.1	0.1	0.1	0.1	0.0
42.1	38.7	40.6	41.6	43.4
0.1	0.1	0.1	0.2	0.3

in the same categories. When the question of increased emphasis is considered, the responses remain rather stable across the college type categories for each issue. The areas of greatest concern to the graduates are preparation in advanced courses and research orientation, as slightly less than one-half of the respondents felt increased emphasis was warranted. A somewhat smaller ratio (40%) indicated encouragement for greater emphasis on student creativity and development of student interest. Approximately one-third of the graduates felt that preparation in laboratory technique,

Preparation in laboratory technique
No response
Strong
Adequate
Weak
No opinion
Invalid
Should be increased
Should be decreased
Preparation in advanced courses
No response
Strong
Adequate
Weak
No opinion
Invalid
Should be increased
Should be decreased

breadth of knowledge, and quality of faculty should be given increased attention in their undergraduate programs. Hence, a substantial percentage of all graduates in all categories believe that these issues should be given increased attention at the undergraduate level.

Comparison of response rates across year of graduation reveals rather stable percentages for the question of preparation in laboratory technique (Table 16.3a). Approximately one-third of the graduates consider laboratory preparation to be strong, and about 40%

TABLE 16.2a

Graduates' Opinions on Adequacy of Types of Training
in Undergraduate Preparation,
by College Type

Male Co-ed N = 7993 %	All-Male N = 3550 %	Female Co-ed N = 3152 %	All-Female N = 1700 %
6.0	6.9	8.7	11.5
35.5	40.0	27.6	30.4
42.6	40.5	39.6	36.1
9.5	7.6	12.7	11.0
6.2	5.0	11.3	11.1
0.0	0.0	0.2	0.0
35.0	33.3	37.7	37.0
2.6	3.9	0.6	1.1
5.5	6.4	7.5	9.2
29.9	37.2	29.6	30.1
43.7	38.7	41.4	41.4
17.4	14.0	17.0	17.0
3.5	3.7	4.3	2.2
0.1	0.1	0.2	0.2
47.8	46.1	47.1	50.9
0.7	0.7	0.3	0.1

Research orientation
  No response
  Strong
  Adequate
  Weak
  No opinion
  Invalid
  Should be increased
  Should be decreased
Quality of faculty
  No response
  Strong
  Adequate
  Weak
  No opinion
  Invalid
  Should be increased
  Should be decreased

of the graduates say it is adequate. One-third of the respondents consistently recommend increased emphasis on laboratory technique in undergraduate studies.

On the other hand, the perceptions of advanced coursework preparation change slightly through the ten-year period considered. In 1958, 29% of the graduates felt that this preparation was strong while 36% had this opinion in 1967. The ratio of graduates that considered this aspect as weak showed a slight decrease for the same period. Yet, rather interestingly, the ratio of graduates within each class recommending strengthened preparation in-

TABLE 16.2b

Graduates' Opinions on Research Orientation and Faculty Quality,
by College Type

Male Co-ed N = 7993 %	All-Male N = 3550 %	Female Co-ed N = 3152 %	All-Female N = 1700 %
5.8	6.6	8.2	9.9
19.9	16.7	19.0	22.2
34.5	37.2	30.1	31.0
33.2	33.8	33.5	30.7
6.5	5.7	9.3	6.1
0.0	0.0	0.0	0.1
44.7	45.9	46.1	46.4
1.9	2.2	0.8	0.5
5.1	5.9	6.8	9.0
57.0	66.6	54.1	52.1
33.4	24.7	34.0	33.8
3.9	2.3	4.4	4.2
0.4	0.4	0.4	0.5
0.2	0.2	0.2	0.4
35.1	32.5	32.1	31.8
0.1	0.0	0.0	0.1

creased from 44% in 1958 to 52% in 1967. Even though more recent graduates felt that their advanced course preparation was stronger than the earlier graduates, they said in increasing numbers that this aspect of undergraduate training warrants increased emphasis.

In 1958, 16% of the graduates felt that research orientation was strong while 24% of the 1967 graduates held the same opinion. Fewer respondents said that this was a strong feature of their undergraduate preparation than in any other area. This is modified slightly by more recent graduates who agree to a larger extent that research orientation received considerable emphasis. Approx-

Breadth of knowledge
No response
Strong
Adequate
Weak
No opinion
Invalid
Should be increased
Should be decreased
Development of student creativity
No response
Strong
Adequate
Weak
No opinion
Invalid
Should be increased
Should be decreased
Development of student interest
No response
Strong
Adequate
Weak
No opinion
Invalid
Should be increased
Should be decreased

TABLE 16.2c

Graduates' Opinions on Breadth of Knowledge and Development
of Student Interest and Creativity,
by College Type

Male Co-ed N = 7993 %	All-Male N = 3550 %	Female Co-ed N = 3152 %	All-Female N = 1700 %
5.4	6.1	7.3	9.3
42.7	48.5	39.0	39.1
44.0	39.0	45.2	42.6
6.9	5.4	7.0	7.5
1.0	0.9	1.5	1.4
0.1	0.0	0.0	0.1
34.6	33.9	32.3	34.4
0.9	0.6	0.2	0.6
5.6	6.3	7.3	9.7
27.2	22.6	23.8	26.5
44.0	46.5	41.8	38.7
18.2	20.3	21.1	20.7
5.0	4.2	6.0	4.3
0.0	0.1	0.1	0.1
42.4	45.8	43.0	41.9
0.3	0.2	0.1	0.1
5.4	6.2	7.4	9.6
36.3	35.0	35.2	39.9
42.2	41.5	41.5	34.8
13.8	14.9	13.0	13.3
2.3	2.4	2.9	2.3
0.1	0.1	0.1	0.1
42.1	43.7	38.1	37.1
0.2	0.1	0.1	0.2

imately one-third of each graduating class stated that this aspect of their preparation was adequate. The percentage of graduates who considered this to be a weak area decreased from 35% in 1958 to 29% in 1967. Of the issues the graduates were asked to respond to, research orientation during undergraduate studies was considered weak by a larger percentage of each class than any other area of this study. The ratio of graduates in each class who had the opinion that increased emphasis is called for approached one-

	1958 N = 1236 %	1959 N = 1292 %	1960 N = 1419 %	1961 N = 1589 %
Preparation in laboratory technique				
No response	7.4	7.3	7.6	6.5
Strong	34.9	35.8	34.2	35.0
Adequate	42.3	39.8	41.7	40.7
Weak	9.6	10.1	9.0	11.2
No opinion	5.8	6.8	7.5	6.6
Invalid	0.1	0.2	0.0	0.0
Should be increased	35.0	33.7	33.8	36.3
Should be decreased	2.1	2.4	2.0	2.3
Advanced courses				
No response	7.0	6.7	7.5	6.4
Strong	28.6	29.8	30.2	29.5
Adequate	41.5	40.7	40.2	42.4
Weak	18.4	18.3	17.9	18.0
No opinion	4.4	4.3	4.2	3.6
Invalid	0.2	0.3	0.0	0.1
Should be increased	43.9	42.7	44.8	47.0
Should be decreased	0.7	0.5	0.4	0.4

half, which indicates considerable dissatisfaction on their part with respect to research orientation in undergraduate preparation. The faculty quality received strong endorsement from all classes with about 60% of the graduates feeling that this was an excellent part of their preparation and another 30% describing it as adequate. Despite this consistent endorsement, more recent graduates stated in increasing numbers that faculty quality should be increased.

TABLE 16.3a

Graduates' Opinions on Adequacy of Types of Training
in Undergraduate Preparation,
by Year

1962 N = 1591 %	1963 N = 1724 %	1964 N = 1800 %	1965 N = 1919 %	1966 N = 1854 %	1967 N = 1942 %
7.7	8.1	7.3	7.0	7.6	6.6
33.6	32.6	34.6	33.3	34.3	36.5
41.6	41.0	39.7	39.8	41.2	41.4
9.5	11.0	9.3	10.9	9.3	8.8
7.6	7.3	8.9	9.0	7.7	6.7
0.1	0.1	0.1	0.0	0.0	0.1
33.3	35.1	34.6	35.6	36.2	38.7
1.9	1.8	3.2	2.0	2.7	2.8
6.9	7.5	6.4	5.7	6.3	5.0
29.6	30.2	32.2	32.5	33.7	35.9
42.7	42.4	42.2	42.5	42.4	41.4
17.9	16.4	15.8	15.7	14.4	14.6
2.9	3.4	3.4	3.6	3.2	3.1
0.0	0.1	0.0	0.1	0.1	0.1
47.4	48.1	48.2	49.0	49.5	51.8
0.4	0.8	0.7	0.7	0.4	0.6

	1958 N = 1236	1959 N = 1292	1960 N = 1419	1961 N = 1589
	%	%	%	%
Research orientation				
No response	7.1	6.3	7.7	6.4
Strong	15.5	15.4	16.3	17.7
Adequate	34.9	35.5	33.8	33.7
Weak	35.4	34.9	34.4	35.3
No opinion	7.0	8.0	7.8	7.0
Invalid	0.1	0.0	0.0	0.0
Should be increased	45.3	44.7	43.6	46.4
Should be decreased	1.7	0.9	1.2	1.2
Quality of faculty				
No response	6.6	6.0	6.9	5.9
Strong	57.4	57.2	55.5	55.8
Adequate	31.6	32.1	33.7	33.0
Weak	3.7	3.9	2.8	4.7
No opinion	0.4	0.4	0.7	0.4
Invalid	0.3	0.4	0.4	0.3
Should be increased	30.0	29.1	31.8	34.4
Should be decreased	0.0	0.1	0.0	0.0

Later graduates recommend strengthening of emphasis on diversity of knowledge, development of student creativity, and development of student interest. Differences in proportions from one class to the next are slight, but are consistent. Creativity was felt to be the weakest area of the three with about one-fifth of the students rating it as weak. It was felt to be a strong area by a smaller percentage of graduates than either diversity or interest. More recently there is a trend towards suggesting increased emphasis on student creativity than in earlier years (42% in 1958 compared to 50% in 1967). Insofar as this study is concerned, the

TABLE 16.3b

Graduates' Opinions on Research Orientation and Faculty Quality,
by Year

1962 N = 1591 %	1963 N = 1724 %	1964 N = 1800 %	1965 N = 1919 %	1966 N = 1854 %	1967 N = 1942 %
8.0	8.0	6.7	6.0	7.0	5.7
17.9	17.2	21.6	21.3	22.3	23.7
33.6	34.3	31.9	33.4	33.0	35.5
33.2	33.2	34.0	32.4	31.7	28.8
7.2	7.2	5.8	7.0	6.0	6.1
0.1	0.0	0.0	0.0	0.0	0.1
42.7	46.2	46.7	45.2	45.5	47.0
1.7	1.6	1.8	1.4	2.1	2.1
6.5	6.8	5.8	5.4	5.9	4.8
56.6	59.1	58.6	58.7	58.8	60.9
32.7	30.6	31.7	30.8	31.3	30.0
3.3	3.0	3.6	4.5	3.5	3.7
0.4	0.5	0.4	0.6	0.4	0.3
0.4	0.0	0.0	0.1	0.1	0.3
31.8	32.1	33.8	34.7	36.7	38.2
0.1	0.2	0.1	0.0	0.1	0.1

respondents indicated a greater need for developing student creativity than either student interest or diversity of knowledge.

Considering breadth of knowledge, increasing percentages of more recent graduates said that this area was strong in their undergraduate preparation (45%) compared to their earlier counterparts (40%). Almost equal percentages in each class rated it as adequate. Even though an increasing ratio of graduates felt that this aspect of undergraduate preparation was strong, they nevertheless recommended further fortification in this area (37% in 1967 compared to 32% in 1958).

	1958 N = 1236 %	1959 N = 1292 %	1960 N = 1419 %	1961 N = 1589 %
Breadth of knowledge				
No response	6.9	6.5	7.3	6.4
Strong	41.2	39.2	40.0	42.2
Adequate	43.1	45.7	44.0	43.1
Weak	7.4	6.9	6.9	7.1
No opinion	1.3	1.6	1.9	1.2
Invalid	0.2	0.1	0.0	0.1
Should be increased	31.6	32.4	34.3	33.1
Should be decreased	0.7	0.5	0.8	0.8
Development of student creativity				
No response	7.0	6.6	7.2	6.5
Strong	22.2	23.8	24.1	26.8
Adequate	44.1	43.3	44.6	41.4
Weak	20.9	21.2	18.7	19.8
No opinion	5.7	5.0	5.4	5.4
Invalid	0.2	0.1	0.0	0.1
Should be increased	41.6	40.5	40.7	42.3
Should be decreased	0.2	0.0	0.4	0.3
Development of student interest				
No response	6.8	6.3	7.2	6.2
Strong	35.0	37.0	36.6	36.5
Adequate	41.8	41.1	40.9	41.5
Weak	13.3	12.9	12.5	13.3
No opinion	3.2	2.8	2.8	2.4
Invalid	0.1	0.0	0.0	0.1
Should be increased	38.2	36.2	38.7	41.1
Should be decreased	0.0	0.0	0.1	0.1

TABLE 16.3c

Graduates' Opinions on Breadth of Knowledge and Development
of Student Interest and Creativity,
by Year

1962 N = 1591 %	1963 N = 1724 %	1964 N = 1800 %	1965 N = 1919 %	1966 N = 1854 %	1967 N = 1942 %
6.9	7.1	6.2	5.6	6.3	5.1
41.3	42.4	44.1	44.5	44.4	46.6
43.8	43.1	42.3	41.8	42.9	41.5
6.8	6.6	6.7	6.9	5.8	6.0
1.2	0.8	0.8	1.2	0.7	1.0
0.1	0.1	0.1	0.0	0.1	0.0
32.1	33.3	34.3	34.2	35.7	37.1
0.8	0.6	0.7	0.5	0.7	0.7
7.0	7.7	6.4	5.9	6.3	5.1
26.7	25.1	26.4	26.1	25.7	26.5
42.5	45.8	43.4	43.0	43.7	43.6
18.4	16.5	19.8	19.7	19.8	20.3
5.3	4.9	3.9	5.4	4.6	4.4
0.1	0.1	0.1	0.0	0.0	0.1
40.4	41.5	42.4	44.7	47.0	47.8
0.2	0.2	0.2	0.2	0.3	0.3
7.1	7.5	6.2	5.7	6.2	5.3
36.0	36.3	38.1	36.1	36.0	34.4
41.6	41.2	39.4	41.8	40.0	42.2
13.1	12.1	14.5	13.3	15.6	16.3
2.1	2.7	1.8	3.2	2.1	1.8
0.1	0.1	0.0	0.0	0.1	0.1
39.4	38.8	40.7	43.0	44.4	47.2
0.2	0.3	0.3	0.1	0.2	0.2

The responses to the question of development of student interest show a marked constancy over the ten-year period, with about 77% rating it as strong or adequate in each class. Approximately 16% of the 1967 graduates said that development of student interest in undergraduate preparation was weak compared to 13% of the earlier graduates. Again, the ratio who felt that increased emphasis is needed rose from 38% in 1958 to 47% in 1967.

The graduates' frequent emphasis on increased college attention to development of student interest and student creativity, and to faculty quality and academic preparation may imply some dissatisfaction, or at least a need for improvement. However, the fact that a large portion of the graduates affirmed strong or adequate training suggests that they are quite willing to endorse and affirm the quality of their training.

Perceptions of competitiveness in postgraduate training may be determined in part by opinions about the adequacy of the preparation received during undergraduate training. In order to appraise this factor, the science graduates who have had postgraduate training were asked to furnish a global judgment about the adequacy of their college education. Data concerning graduates with advanced training are given in Tables 16.4, 16.5 and 16.6. The low percentage of graduates who would not or could not form judgments is given in the category of "No Rank" in each table.

Of the judgmental choices available—excellent, average, poor, and no rank—the respondents were most likely to describe their own training as excellent. Examining the percentages in Table 16.4, we find that the modal category of endorsement is the excellent rating for each major discipline except physics; the latter major is somewhat more critical and more frequently provides a rating of average to their undergraduate training. The pre-med graduates were most likely to endorse their training as excellent compared to the other subject areas. From Table 16.5, it is seen that graduates of men's colleges are most likely to endorse the quality of their preparation and the graduates of women's colleges

are least likely to do so. Low opinions of their own college preparation for graduate school were expressed by only a small percentage
(3–9%) of respondents regardless of their major discipline, type of
college, or year of graduation. When the graduates' opinions are
examined by year of graduation (Table 16.6), two trends are apparent: the more recent graduates endorse their preparation as
excellent more frequently than the earlier graduates, and the more
recent graduates less frequently feel that their undergraduate
preparation was poor.

Graduates also judged the excellence of training afforded by
other kinds of institutions: other liberal arts colleges, prestigious
private and state universities, and leading technical schools. The
results of these objective judgments, based upon the experience
of science graduates who have had advanced training, are also
given in Tables 16.4, 16.5, and 16.6. In evaluating the training of
other liberal arts colleges, graduates were most likely to say that
it is average. Graduates of men's colleges rendered a high opinion
more frequently than graduates of women's colleges. Also, the
highest proportion of excellent ratings were awarded to other liberal arts institutions by biology, chemistry, and pre-med graduates.
There are no substantial trends indicated when the graduates'
opinions by year of graduation are considered.

Graduates with postgraduate education rated the prestigious
state universities in a similar fashion as the other liberal arts colleges. No substantial differences are evident and the modal rating
is average even though excellent ratings are more frequent. Graduates with postgraduate education rated the prestigious private
universities and leading technical schools excellent almost as frequently as they did their own undergraduate education. The technical institutions were most similar to the graduates' own institutions in this regard; preparation by private universities was considered average by a higher proportion of graduates.

When judgments about the substantive areas of background
in humanities and facility with arts of communication are made,

My college
  No response
  Excellent
  Average
  Poor
  No rank
Other liberal arts colleges
  No response
  Excellent
  Average
  Poor
  No rank
Prestigious private universities
  No response
  Excellent
  Average
  Poor
  No rank
Prestigious state universities
  No response
  Excellent
  Average
  Poor
  No rank
Leading technical schools
  No response
  Excellent
  Average
  Poor
  No rank

TABLE 16.4

Graduates' Opinions of Their Undergraduate Preparation for Graduate
School in Comparison to Other Types of Institutions,
by Major Subject*

Biology N = 3960 %	Chemistry N = 2522 %	Math N = 2126 %	Physics N = 1532 %	Pre-med N = 1059 %
5.3	7.7	10.0	13.6	3.9
58.6	51.4	40.8	33.5	65.0
31.0	34.9	37.6	40.5	27.9
2.9	3.8	6.4	9.0	1.6
2.2	2.2	5.2	3.4	1.6
13.7	14.3	17.4	19.5	11.0
22.4	19.2	13.6	10.9	23.0
52.9	52.9	48.7	48.9	57.8
2.4	4.5	5.6	7.2	1.6
8.4	9.1	14.7	13.5	6.6
14.6	14.8	18.2	20.8	12.1
38.9	39.5	34.5	33.4	43.1
30.9	29.0	24.7	24.7	32.8
2.3	1.8	1.8	2.5	1.4
13.3	14.9	20.8	18.6	10.6
14.8	14.7	17.9	20.4	12.2
22.5	25.7	22.0	21.9	21.6
44.4	41.0	36.7	36.2	50.2
5.8	4.9	4.6	4.5	5.6
12.5	13.7	18.8	17.0	10.4
16.4	15.5	17.6	19.0	14.8
39.7	49.8	44.6	51.3	41.3
15.2	14.1	12.9	11.8	16.7
2.8	2.9	3.1	4.8	2.6
25.9	17.7	21.8	13.1	24.6

*Note: Respondents to this question were those who were enrolled in graduate
school for any period of time.

My college
  No response
  Excellent
  Average
  Poor
  No rank
Other liberal arts colleges
  No response
  Excellent
  Average
  Poor
  No rank
Prestigious private universities
  No response
  Excellent
  Average
  Poor
  No rank
Prestigious state universities
  No response
  Excellent
  Average
  Poor
  No rank
Leading technical schools
  No response
  Excellent
  Average
  Poor
  No rank

TABLE 16.5

Graduates' Opinions of Their Undergraduate Preparation for Graduate
School in Comparison to Other Types of Institutions,
by College Type*

Male Co-ed N = 6210 %	All-Male N = 2670 %	Female Co-ed N = 1548 %	All-Female N = 879 %
11.6	5.5	2.8	7.5
46.4	61.7	50.5	43.3
34.5	28.3	37.9	38.9
4.6	2.8	4.3	6.9
2.9	1.7	4.5	3.4
18.2	11.4	13.2	17.6
17.8	22.4	16.3	13.9
49.9	56.1	53.0	48.2
4.2	2.9	3.9	5.9
9.9	7.2	13.6	14.4
19.0	12.2	14.7	18.5
34.6	43.9	34.9	42.7
28.8	31.8	24.0	21.7
2.2	1.3	2.7	1.8
15.4	10.8	23.7	15.3
19.0	12.2	14.0	18.6
22.3	22.8	22.8	25.9
40.5	48.2	37.5	32.3
4.9	5.5	5.3	4.0
13.3	11.3	20.4	19.2
19.5	14.2	15.3	17.3
42.1	47.3	42.6	53.6
14.8	16.8	9.4	7.7
3.2	2.2	3.5	3.6
20.4	19.5	29.2	17.8

*Note: Respondents to this question were those who were enrolled in graduate
school for any period of time.

	1958 N = 837 %	1959 N = 884 %	1960 N = 939 %	1961 N = 1088 %
My college				
No response	7.1	8.6	5.6	7.2
Excellent	48.5	48.9	50.1	49.8
Average	35.5	33.7	35.2	35.5
Poor	6.7	6.6	5.6	4.6
No rank	2.2	2.2	3.5	2.9
Other liberal arts colleges				
No response	18.6	19.4	15.2	14.2
Excellent	17.2	17.1	19.6	16.7
Average	49.9	49.5	52.4	53.8
Poor	4.7	4.3	3.1	4.9
No rank	9.6	9.7	9.7	10.4
Prestigious private universities				
No response	18.9	19.6	16.4	15.1
Excellent	34.1	36.1	38.6	38.3
Average	30.1	28.0	27.4	29.9
Poor	2.8	2.7	1.4	2.1
No rank	14.1	13.6	16.2	14.6
Prestigious state universities				
No response	18.8	19.1	16.2	14.9
Excellent	21.3	20.7	23.6	24.9
Average	40.7	39.5	41.5	42.7
Poor	5.7	7.0	4.8	4.2
No rank	13.5	13.7	13.9	13.3
Leading technical schools				
No response	19.1	19.9	16.9	15.9
Excellent	45.9	44.2	46.5	47.5
Average	13.3	14.4	13.3	13.3
Poor	3.0	2.6	3.5	3.7
No rank	18.7	18.9	19.8	19.6

TABLE 16.6

Graduates' Opinions of Their Undergraduate Preparation for Graduate School in Comparison to Other Types of Institutions, by Year*

1962 N = 1104 %	1963 N = 1227 %	1964 N = 1273 %	1965 N = 1334 %	1966 N = 1251 %	1967 N = 1249 %
9.0	10.6	8.7	8.5	8.1	3.9
46.5	49.3	51.1	50.1	53.9	57.9
36.9	33.3	33.7	33.3	32.4	33.1
4.5	3.7	4.2	4.5	2.9	2.8
3.1	3.1	2.3	3.6	2.7	2.3
16.0	18.2	14.7	14.4	13.8	9.3
18.2	18.6	18.7	17.6	19.7	20.7
51.1	51.6	51.9	52.3	52.6	54.7
3.9	3.1	4.5	4.7	2.8	3.9
10.8	8.5	10.2	11.0	11.1	11.4
16.8	18.8	15.5	15.3	15.0	10.7
38.0	38.3	39.5	36.1	38.3	39.8
29.0	26.7	27.2	28.9	29.1	29.9
2.2	1.7	2.0	1.9	1.8	1.8
14.0	14.5	15.8	17.8	15.8	17.8
16.4	19.2	15.4	15.5	15.7	9.9
24.9	22.5	23.7	20.9	22.3	24.8
39.4	39.4	41.5	42.2	42.8	45.5
5.5	4.0	5.6	4.7	4.7	5.1
13.8	14.9	13.8	16.7	14.5	14.7
17.1	19.3	16.5	16.4	16.8	10.8
44.2	44.2	43.6	42.3	42.7	47.0
14.3	13.3	14.3	14.4	15.9	14.5
4.1	3.1	3.4	2.9	2.9	2.2
20.3	20.1	22.2	24.0	21.7	25.5

*Note: Respondents to this question were those who were enrolled in graduate school for any period of time.

the evaluations are in high agreement (Tables 16.7 and 16.8). Approximately 80% of the graduates felt that they had received a broad background in humanities. The male college graduate was less likely to say that he had received a broad background in humanities than the female graduate. The female college graduate was least likely to say her background in humanities was inadequate. The responses indicate that most graduates are satisfied with the humanities background that was provided in their undergraduate programs.

Considering the question of whether or not the baccalaureate program provided them with facility in the art of communication, again most respondents felt that such was the case. This held true regardless of whether comparisons were made by subject, college type, or year of graduation with about 73% affirming a positive response. However, a slightly larger percentage (14%) provided negative response than in the question of humanities

Did undergraduate years provide:

Broad background in humanities?
  No response
  Yes
  No
Facility with arts of communication?
  No response
  Yes
  No

background. Nevertheless, there appears to be substantial endorsement by the graduates of all colleges providing in their programs the necessary framework for the graduates to acquire the art of communication with others.

In Tables 16.9, 16.10, and 16.11, the graduates' perceptions of competitiveness in graduate studies as opposed to undergraduate training are examined. Among the science graduates, physics majors were most likely to hold the opinion that graduate school is more competitive than their undergraduate experience. Biology majors were least likely to say this is true. Chemistry, math and pre-med graduates are quite similar in their responses, and the frequency with which these groups endorse the contention that graduate programs are more competitive is intermediate between the biology and physics levels of response. More than one-third of each group saw no major differences in competitiveness at the undergraduate and graduate levels.

TABLE 16.7

Graduates' Opinions on Undergraduate Preparation as Providing
Background in Humanities and Communication,
by College Type

Male Co-ed N = 7993 %	All-Male N = 3550 %	Female Co-ed N = 3152 %	All-Female N = 1700 %
13.3	14.2	11.2	13.4
78.4	75.9	82.3	81.7
8.3	9.8	6.5	4.9
14.0	14.8	12.3	14.5
72.3	72.6	72.3	73.7
13.7	12.6	15.4	11.8

Did undergraduate years provide:	1958 N = 1236 %	1959 N = 1292 %	1960 N = 1419 %	1961 N = 1589 %
Broad background in humanities?				
No response	14.1	12.7	13.5	12.2
Yes	78.6	79.4	79.1	80.8
No	7.3	7.9	7.4	7.0
Facility with arts of communication?				
No response	15.1	13.7	14.2	13.0
Yes	70.9	71.8	71.7	72.7
No	14.0	14.4	14.2	14.3

TABLE 16.9

Graduates' Opinions on Competition in Graduate School, by Major Subject*

	Biology N = 3957 %	Chemistry N = 2521 %	Math N = 2126 %	Physics N = 1532 %	Pre-med N = 1059 %
Less imposing than at college	37.1	30.3	29.9	23.6	26.4
More imposing than at college	29.0	32.4	35.1	38.7	34.1
About the same as at college	34.0	37.3	34.8	37.7	39.6

*Note: It was required that those answering this question have attended graduate school for at least one semester.

TABLE 16.8

Graduates' Opinions on Undergraduate Preparation as Providing
Background in Humanities and Communication,
by Year

1962 N = 1591 %	1963 N = 1724 %	1964 N = 1800 %	1965 N = 1919 %	1966 N = 1854 %	1967 N = 1942 %
13.5	13.9	12.7	13.1	13.8	12.0
78.3	79.1	78.3	78.5	77.5	80.2
8.2	7.1	8.9	8.4	8.7	7.9
14.3	14.9	13.2	13.9	14.5	12.7
73.2	71.5	73.5	72.6	71.4	74.5
12.5	13.6	13.3	13.4	14.0	12.9

TABLE 16.10

Graduates' Opinions on Competition in Graduate School,
by College Type*

	Male Co-ed N = 6011 %	All-Male N = 2758 %	Female Co-ed N = 1549 %	All-Female N = 879 %
Less imposing than at college	27.5	30.9	40.8	41.4
More imposing than at college	36.0	33.5	22.0	27.3
About the same as at college	36.4	35.6	37.2	31.2

*Note: It was required that those answering this question have attended graduate
school for at least one semester.

	1958 N = 826 %	1959 N = 883 %	1960 N = 939 %	1961 N = 1088 %
Less imposing than at college	25.1	27.3	26.9	27.6
More imposing than at college	36.8	35.3	35.8	34.2
About the same as at college	38.1	37.3	37.3	38.0

The male co-ed graduate is as likely to say that graduate school is more competitive as he is to say that it is about the same as undergraduate school. Among all female graduates, there is a definite opinion that graduate training is less competitive than undergraduate school. Sex differences in the rate at which graduates enter postgraduate programs and subject areas may account for this difference in opinion about the competitiveness of graduate training between college types, but further research would be necessary to affirm this statement. Approximately one-third of each category felt that there is no major difference between competitiveness in their baccalaureate and post-baccaluareate programs.

When the data are classified by year of graduation, slight trends are evident. In general, it appears that the ratio of individuals who feel that graduate requirements are less exacting than the undergraduate is increasing (from 25% to 35% during 1958–1967). More competitiveness is perceived by the early graduates (37% of the 1958 class) than by later graduates (32% of the 1967 class). The changes are fairly stable in that increases (or decreases) of 1 to 2% are apparent in each graduating class.

When graduates were asked if they would choose a liberal arts

TABLE 16.11

Graduates' Opinions on Competition in Graduate School,
by Year*

1962 $N = 1104$ %	1963 $N = 1227$ %	1964 $N = 1273$ %	1965 $N = 1334$ %	1966 $N = 1252$ %	1967 $N = 1249$ %
31.7	32.3	32.5	34.4	36.2	34.6
32.9	32.4	29.3	30.9	30.8	31.8
35.2	35.3	38.1	34.7	33.0	33.6

*Note: It was required that those answering this question have attended graduate school for at least one semester.

college again if they had the opportunity, the overwhelming majority (90%) of graduates said they would re-select this type of institution (Tables 16.12, 16.13, and 16.14). From 4 to 7% of each group would not do so, however, and about 5% of the graduates provided no response to the inquiry. Relative to other major disciplines, physics majors appear to be the most dissatisfied with about 7% saying they would not select a liberal arts college again. Approximately 4% of the graduates in the other disciplines would not attend a liberal arts college again if given another opportunity.

With regard to college types, the male co-ed graduates signaled the most dissatisfaction with over 5% saying they would not select a liberal arts college again. The female college graduate was least likely to hold the opinion that she would not attend a liberal arts college given the opportunity again. When year of graduation is considered, no trends are evident. Nearly 90% of the graduates in each class affirm the choice of a liberal arts college again and 4 to 5% state that they would not make this selection again. From the responses, it is evident that the vast majority of graduates are satisfied with their choice of a liberal arts college for their education.

TABLE 16.12

Graduates' Retrospective Views on Choice of a Liberal Arts College,
by Major Subject

If you could choose again, would you choose a liberal arts college?	Biology N = 5996 %	Chemistry N = 3362 %	Math N = 3666 %	Physics N = 2045 %	Pre-med N = 1329 %
No response	7.1	6.0	5.3	6.1	5.9
Yes	88.9	89.9	90.1	87.4	90.1
No	4.0	4.0	4.6	6.5	3.8

SUMMARY

The most frequent graduate judgments are largely positive rather than negative for a variety of undergraduate training characteristics. The majority of graduates declared that laboratory training, advanced coursework, and faculty quality and other attributes were strong or adequate. The most evident approval was demon-

If you could choose again would you choose a liberal arts college?	1958 N = 1236 %	1959 N = 1292 %	1960 N = 1419 %	1961 N = 1589 %
No response	6.6	6.0	6.6	6.2
Yes	88.9	90.1	88.9	88.7
No	4.5	4.0	4.4	5.0

TABLE 16.13

Graduates' Retrospective Views on Choice of a Liberal Arts College,
by College Type

If you could choose again would you choose a liberal arts college?	Male Co-ed N = 7993 %	All-Male N = 3550 %	Female Co-ed N = 3152 %	All-Female N = 1700 %
No response	5.5	6.1	6.9	8.8
Yes	89.1	89.8	89.6	88.4
No	5.3	4.1	3.5	2.8

strated by responses to the inquiry about faculty quality. More than half of the respondents said that this was a strong rather than adequate or weak feature of their education. On the other hand, the undergraduate emphasis on research orientation was most likely to be acknowledged as weak relative to other college characteristics.

TABLE 16.14

Graduates' Retrospective Views on Choice of a Liberal Arts College,
by Year

1962 N = 1591 %	1963 N = 1724 %	1964 N = 1800 %	1965 N = 1919 %	1966 N = 1854 %	1967 N = 1942 %
6.9	7.0	5.9	5.7	6.3	5.6
88.2	89.3	89.5	89.6	89.2	90.3
4.8	3.8	4.6	4.6	4.5	4.1

A comparison of the data for major discipline categories reveals that pre-med graduates are most likely to affirm strong attributes of their undergraduate training. In regard to the college's emphasis on development of student creativity and interest, and research orientation, however, these graduates give fewer positive replies than do members of the other major disciplines. Physics graduates were somewhat more likely than other respondents to view undergraduate emphasis on advanced coursework, laboratory techniques, faculty quality, and breadth of knowledge as being weak.

A substantial number of the respondents expressed the opinion that undergraduate attention to all features should be increased, particularly for the need of increased advanced coursework. Fewer graduates recommended more attention to faculty quality and breadth of knowledge than to the other features of this study.

In general, few graduates found undergraduate training to be inadequate with respect to the criteria examined here. Considerable numbers recommended increased college attention to the qualitative aspects of their training. It is likely that many of these respondents are also those who felt the educational emphasis to be adequate rather than strong. Attention to the development of student interest and creativity, and advanced coursework are most likely to be perceived as being something less than adequate, but the proportion of graduates having this opinion depends on the major discipline.

The graduates, whether they had had post-baccalaureate training or not, felt that their college had provided them with excellent training with few saying that it was poor. When asked to evaluate other liberal arts colleges, the same graduates felt that the training provided here was adequate with few saying that it was on the same level as provided by their own institutions. The leading technical schools won endorsement from graduates who had post-baccalaureate training, but did not do so among those graduates who had not. The prestigious private universities were felt by many to provide a better undergraduate preparation for both graduate school and life outside college than the state universities in this study.

A large majority of the graduates endorsed the background in humanities and facility with arts of communication that they had received at their institutions. From the responses, the graduates sense that the colleges are providing an excellent framework in these areas to be used later in their careers.

When the graduates were asked their perceptions on competitiveness in graduate studies as opposed to undergraduate studies, approximately equal proportions felt that graduate requirements were more demanding, less demanding or equally imposing. Two trends are evident: female graduates hold the opinion that competitiveness in graduate school is less imposing than in undergraduate school, and more recent graduates perceive that graduate school is less competitive than earlier graduates.

The overwhelming majority of graduates, even though they recognize certain weaknesses in their undergraduate programs, endorse the quality of training provided by the liberal arts college and would re-select the same college for their baccalaureate program if given the opportunity.

Selection Process and List
of Participating Colleges

The population of colleges which were of basic interest to Research Corporation when this study was undertaken in 1967 included 150 liberal arts institutions with strong undergraduate programs in the sciences. Each of the colleges was proposed by one or more of the foundation's staff, based on firsthand knowledge of the institutions. This initial pool consisted of institutions ranging from the nationally known and popularly recognized to those with local or regional reputations for excellence.

The selections were necessarily subjective, being based on opinions as to overall educational quality and the caliber of science programs. The administrative and sampling criteria used in determination of the final study group included the desirability of an adequate geographic cross section of colleges, and the availability of normative performance records on students. The latter data comprise Scholastic Aptitude Test scores administered nationally by the College Entrance Examination Board to students prior to college entry.

For economy of sampling, institutional participation was restricted to a total of 50 institutions—10 men's, 10 women's colleges and 30 coeducational colleges. One of the women's colleges was unable to provide the necessary student data and was eliminated, with no substitute being provided. The final study group, therefore, is made up of 49 colleges.

Four additional colleges were chosen from the initial pool as a test sample for a preliminary evaluation of the data-gathering materials and techniques. The experience gained in the test survey made it possible to go on with the large study, but since it resulted in changes in questionnaires and data forms, the findings of that survey are not included with those of the whole study.

Following are the colleges studied in the test survey, the findings for which are *not* included here:

Colorado College	Juniata College
Haverford College	Willamette University

Following are the 49 colleges which participated in the full study and for which the findings are presented here:

Allegheny College	Gettysburg College	Reed College
Antioch College	Grinnell College	Rollins College
Bates College	Hamilton College	St. Lawrence
Beloit College	Hobart and William	University
Bennington College	Smith Colleges	Swarthmore College
Bowdoin College	Kalamazoo College	Sweet Briar College
Bryn Mawr College	Kenyon College	Trinity College (Conn.)
Carleton College	Knox College	University of the South
Clarkson College of	Lake Forest College	Vassar College
Technology	Lewis and Clark	Wabash College
Colby College	College	Washington and Lee
Connecticut College	Middlebury College	College
Davidson College	Mills College	Wellesley College
Denison University	Muhlenburg College	Wheaton College
DePauw University	Occidental College	(Mass.)
Dickinson College	Ohio Wesleyan	Whitman College
Franklin and Marshall	University	Williams College
College	Pomona College	Wilson College
Furman University	University of Redlands	College of Wooster

Admissions Office Form,
Entering Science Student Data

RESEARCH CORPORATION: Study of Liberal Arts Science

ENTERING SCIENCE STUDENT DATA

ADMISSIONS OFFICE

COLLEGE _____

STUDENT'S NAME _____

NAME OF HIGH SCHOOL (PRINT IN BOX BELOW)

SEX: MALE ::::::   FEMALE ::::::

SPECIAL SCIENCE & MATH. HONORS

WESTINGHOUSE ::::::
BAUSCH & LOMB ::::::
AREA SCIENCE FAIR ::::::
N.S.F. INSTITUTE ::::::
HIGH SCHOOL AWARD ::::::
OTHER ::::::

IDENTIFICATION NUMBER

0	1	2	3	4	5	6	7	8	9
0	1	2	3	4	5	6	7	8	9
0	1	2	3	4	5	6	7	8	9
0	1	2	3	4	5	6	7	8	9
0	1	2	3	4	5	6	7	8	9
0	1	2	3	4	5	6	7	8	9
0	1	2	3	4	5	6	7	8	9
0	1	2	3	4	5	6	7	8	9

SCHOLARSHIP ASSISTANCE AT ENTRANCE:

GRANTED BY: GOVERNMENT ::::   COLLEGE ::::   PRIVATE SOURCE ::::

GRANTED FOR: GEN. SCHOLASTIC ACHIEVEMENTS ::::   SCIENCE PROWESS ::::   ECONOMIC NEED ::::   OTHER SKILLS OR ACHIEVEMENTS ::::

PROBABLE MAJOR AT ENTRANCE:

PHYSICS ::::::
CHEMISTRY ::::::
PRE-MED ::::::
BIOLOGY ::::::
MATHEMATICS ::::::

YEAR OF ENTRY AT COLLEGE: (SAMPLE 1956)

0	1	2	3	4	5	6	7	8	9
					5				
0	1	2	3	4	5	6	7	8	9
						6			

ADVANCED PLACEMENT AND/OR CREDIT:

BIOLOGY ▓ PHYSICS ▓

MATHEMATICS ▓ CHEMISTRY ▓

TOTAL NUMBER OF HIGH SCHOOL SCIENCE & MATH COURSES

0	1	2	3	4	5	6	7	8	9
10	11	12	13	14	15	16	17	18	19

NUMBER OF HIGH SCHOOL SCIENCE & MATH HONOR COURSES:

0	1	2	3	4	5	6	7	8	9+

HIGH SCHOOL SCIENCE & MATH AVERAGE:

A+	A-	B+	B	B-	C+	C	C-	D+	D

CHARACTERISTICS OF HIGH SCHOOL OF ORIGIN

(1) LARGE	MEDIUM	SMALL
(2) URBAN	SUBURBAN	RURAL
(3) HIGHLY ACADEMIC	AVERAGE ACADEMIC	WEAK ACADEMIC
(4) STRONG SCIENCE	AVERAGE SCIENCE	WEAK SCIENCE
(5) LARGELY COLLEGE PREP	20%-50% COLLEGE PREP	LARGELY VOCATIONAL

COLLEGE BOARD–SCHOLASTIC APTITUDE TEST:

SAMPLE: SCORE 469

(ENTER SCORES IN BOXES FROM TOP TO BOTTOM)

| 4 | 6 | 9 |

VERBAL

0	1	2	3	4	5	6	7	8	9
0	1	2	3	4	5	6	7	8	9
0	1	2	3	4	5	6	7	8	9

QUANTITATIVE

0	1	2	3	4	5	6	7	8	9
0	1	2	3	4	5	6	7	8	9
0	1	2	3	4	5	6	7	8	9

9S696H IBM

Registrars Office Form, Science Graduate Information Sheet

RESEARCH CORPORATION: Study of Liberal Arts Science

SCIENCE GRADUATE INFORMATION SHEET

REGISTRARS OFFICE

COLLEGE _____

GRADUATE'S NAME _____

DIRECTIONS: Read each statement and its lettered answers. When you have decided on your answer, blacken the corresponding space with a **NO. 2 PENCIL.** Make your mark as long as the pair of lines, and completely fill the area between the pair of lines. If you change your mind, erase your first mark COMPLETELY. Make no stray marks.

IDENTIFICATION NUMBER

	0	1	2	3	4	5	6	7	8	9

APPROXIMATE SIZE OF COLLEGE CLASS:

0-40	241-280
41-80	281-320
81-120	321-360
121-160	361-400
161-200	401-440
201-240	441-480 & UP

APPROXIMATE RANK IN COLLEGE CLASS:

1- 5% (TOP)	51-60%
6-10%	61-70%
11-20%	71-80%
21-30%	81-90%
31-40%	91-95%
41-50%	96-100% (BOTTOM)

GRADUATE RECORD EXAMINATIONS:

SAMPLE (ENTER SCORES IN BOXES FROM TOP TO BOTTOM) ie SCORE—324

		3
		2
		4

MAJOR (MUST BE ONE OF THE FOLLOWING)

BIOLOGY   CHEMISTRY   PRE-MED

PHYSICS

MATHEMATICS

THIS SPACE

‖‖ ‖‖ ‖‖ ‖‖ ‖‖ ‖‖ ‖‖ ‖‖ ‖‖ ‖‖ ‖‖ ‖‖ ‖‖ ‖‖ ‖‖ ‖‖

APTITUDE (QUANTATIVE ONLY)

0	1	2	3	4
0	1	2	3	4
0	1	2	3	4

ADVANCED (ACHIEVEMENT TEST IN MAJOR)

0	1	2	3	4	5	6	7	8	9
0	1	2	3	4	5	6	7	8	9
0	1	2	3	4	5	6	7	8	9

GRADE POINT AVERAGES:

	OVERALL										MAJOR AREA ONLY				
4.00-3.75 (A or A+)															
3.74-3.25 (A– or B+)															
3.24-2.75 (B)															
2.74-2.25 (B– or C+)															
2.24-1.75 (C)															
1.74-1.25 (C– or D+)															
1.24– 0 (D or less)															

HONOR SOCIETIES:

PHI BETA KAPPA		PHI LAMBDA UPSILON
PHI KAPPA PHI		PI MU EPSILON
ALPHA EPSILON DELTA		BETA BETA BETA
SIGMA PI SIGMA		OTHER

EARNED COLLEGE HONORS:

EARNED DEPARTMENT HONORS:

CUM LAUDE	HIGHEST HONORS
SUMMA CUM LAUDE	HIGH HONORS
MAGNA CUM LAUDE	HONORS

IBM H96512

253

Graduate Questionnaire

RESEARCH CORPORATION: Study of Liberal Arts Science

A Foundation

GRADUATE QUESTIONNAIRE

SAM S. SMITH
5064301

NOTE: All of the information is to be coded and used in group analysis for the study purposes only; your responses will be held in strictest professional confidence.

DIRECTIONS: Your responses will be read by an automatic scanning device. Your careful observance of these few simple rules will be most appreciated.

Use only black lead pencil (No. 2½ or softer).

Erase cleanly any answer you wish to change.

Make heavy black marks that fill the circle.

Make no stray markings of any kind.

EXAMPLE: Will marks made with ball pen or fountain pen be properly read? ....

YES ○   NO ●

SECTION A.   YOUR PRE-COLLEGE AND COLLEGE YEARS.

1. At approximately what grade level did you decide:

	PRIOR TO GRADE 9		HIGH SCHOOL			COLLEGE +					
		9	10	11	12	I	II	III	IV	V	VI
that science was a field of interest to you?	○	○	○	○	○	○	○	○	○	○	○
to seek the college major in which you graduated?	○	○	○	○	○	○	○	○	○	○	○
what your highest degree was to be?	○										

2. What was the influence of each of the following upon your decision to apply to, and enter, a liberal arts college?

	A DETERRENT	NO INFLUENCE	MILD INFLUENCE	STRONG INFLUENCE
membership in a relatively small student body.	○	○	○	○
the flexibility of a liberal arts curriculum.	○	○	○	○
"small class and close faculty-student ties."	○	○	○	○
membership in activities (such as athletics, band, or dramatics) that were less competitive than at larger schools.	○	○	○	○
the broad background offered by the liberal arts.	○	○	○	○
relatives attended the college.	○	○	○	○
close friends were attending the college.	○	○	○	○
the college was near at hand.	○	○	○	○
the college was far from home.	○	○	○	○
a scholarship was granted.	○	○	○	○
the college was noted for strength in math/science.	○	○	○	○
admission was assured.	○	○	○	○
impression created by college facilities.	○	○	○	○
impression created by science facilities.	○	○	○	○
promotion of alumni.	○	○	○	○
members of opposite sex were close at hand.	○	○	○	○
recommendation of school counselor to attend.	○	○	○	○
recommendation of high school science teacher.	○	○	○	○
parents wanted a liberal arts college.	○	○	○	○

257

3. What was the effect of each of the following upon your decision to major in the area that you did?

	STRONG INFLUENCE	MILD INFLUENCE	NO INFLUENCE	A DETERRENT
previous success in basic courses.	O	O	O	O
empathy toward faculty in the department.	O	O	O	O
challenge of the curriculum.	O	O	O	O
facilities of the department.	O	O	O	O
quality of the faculty.	O	O	O	O
offered direct route to graduate or professional school goal.	O	O	O	O
led immediately to a job with good compensation.	O	O	O	O
aspirations of relatives or friends.	O	O	O	O

4. Indicate, in terms of ten percents, the contribution of each of the following sources to your undergraduate expenses.

TEN PERCENTS

	0	1	2	3	4	5	6	7	8	9+
Parental or other family aid.	O	O	O	O	O	O	O	O	O	O
Repayable loans.	O	O	O	O	O	O	O	O	O	O
Scholarships, grants, or other gifts.	O	O	O	O	O	O	O	O	O	O
Work during school year.	O	O	O	O	O	O	O	O	O	O
Summer work.	O	O	O	O	O	O	O	O	O	O
Personal savings.	O	O	O	O	O	O	O	O	O	O
Spouse's earnings.	O	O	O	O	O	O	O	O	O	O

5. What is your best estimate of your parents' yearly income at the time you entered college?

Under $7,000 ...... O $13,000 – $18,999 .. O   Over

$7,000 – $12,999 .. O $19,000 – $24,999 .. O $25,000 .. O

SECTION B. YOUR GRADUATE SCHOOL YEARS. (If you have not attended graduate school, go on to Section C)

1. What was your score on the aptitude test of the Graduate Record Exam?

[ ][ ][ ]   0 O 1 O 2 O 3 O 4 O 5 O 6 O 7 O 8 O 9 O

I did not take the exam. .... O

I can not recall. .... O

2. To which graduate schools did you apply, at which were you accepted, and which did you attend? Find your school(s) on the list provided on the instruction sheet and enter each number appropriately.

SCHOOL

YEARS ATTENDED

ATTENDED: Part Time / Full Time

Applied | Accepted | 58/59 | 59/60 | 60/61 | 61/62 | 62/63 | 63/64 | 64/65 | 65/66 | 66/67

OTHER:

OPEN TO PAGE 3

2.

3. Which of the following most closely identifies your area of specialization?

biology.............. ○
chemistry.............. ○
engineering.............. ○
health professions.............. ○
mathematics.............. ○
physics.............. ○
interdisciplinary science.......... ○
other physical sciences.......... ○
agriculture.............. ○
architecture.............. ○
business and commerce.......... ○
educator.............. ○

English and journalism... ○
fine and applied arts ..... ○
foreign languages and lit. ○
forestry .... ○
geography .... ○
home economics........... ○
law .... ○
library science .... ○
mathematical subjects ... ○
philosophy .... ○
psychology .... ○
religion .... ○
social sciences .... ○

4. What graduate degrees have you received and/or which do you intend to receive?

	Received	Intend to Receive
Master's degree (M.A., M.S., etc.)	○	○
Doctor's degree (Ph.D., Ed.D., etc.)	○	○
Medical degree (M.D., D.D.S., D.V.M., etc.)	○	○
Law degree (Ll.B., J.D.)	○	○
Divinity degree (B.D.)	○	○
Other	○	○

8. Please identify the source of any financial assistance (fellowships, assistantships, etc.) which you received for your graduate or professional education.

SOURCE OF STIPEND	APPLIED FOR (or was nominated)	AWARD NOT OFFERED	AWARD ACCEPTED	REFUSED
a. Federal government				
1) Atomic Energy Commission	○	○	○	○
2) Department of Defense	○	○	○	○
3) National Science Foundation	○	○	○	○
4) Veteran's Administration	○	○	○	○
5) National Aeronautics and Space Administration	○	○	○	○
6) U.S. Office of Education:				
National Defense Ed. Act	○	○	○	○
Other Office of Education	○	○	○	○
7) U.S. Public Health Service:				
N.I.H. Fellowship Program	○	○	○	○
N.I.H. Trainee Program	○	○	○	○
Other Public Health Service	○	○	○	○
8) Other Federal Government	○	○	○	○
b. Woodrow Wilson	○	○	○	○
c. Fulbright	○	○	○	○
d. Other private source	○	○	○	○
e. Directly from the school I attended (am attending)	○	○	○	○
f. Other	○	○	○	○

5. Please estimate your average within graduate school.

A or A+ ○ A- ○ B+ ○ B ○ B- ○ C+ ○ C ○ C- ○ D ○

6. In terms of semester-hours, approximately how many hours have you completed in graduate study? (Include any in which you are now enrolled.)

20 or less ○ 25 ○ 30 ○ 35 ○ 40 ○ 45 ○ 50 ○ 55 or more ○

7. Once having completed your highest degree, do you intend to go further? (or have you gone further?)

Specialty (non-degree) courses ○ post doctoral study ○ pursuit of a medical specialty ○

---

9. Which of the following best describes the type of stipend that you held (or now hold)?

Teaching assistantship ○ Research assistantship ○

Work free stipend (equal to or less than tuition) ○

Work free stipend (tuition plus cash grant) ○ No stipend awarded ○

10. What was (is) your principal source of income within graduate school?

	NOT A SOURCE	(1–25%)	(26%–50%)	MAJOR SOURCE (over 50%)
Parents	○	○	○	○
Fellowships	○	○	○	○
Assistantships	○	○	○	○
Loans	○	○	○	○
Personal savings	○	○	○	○
Part time work	○	○	○	○
Spouse's earnings	○	○	○	○

SECTION C. PERSONAL HISTORY.

1. During which of the following years have you been married? (Mark all that apply)

54 ○ 55 ○ 56 ○ 57 ○ 58 ○ 59 ○ 60 ○ 61 ○ 62 ○ 63 ○ 64 ○ 65 ○ 66 ○ 67 ○

2. With respect to each of the following, indicate the year or years that apply.

	58	59	60	61	62	63	64	65	66	67
Graduation from college	○	○	○	○	○	○	○	○	○	○
Years spent in armed services	○	○	○	○	○	○	○	○	○	○
Years in Peace Corps (or similar)	○	○	○	○	○	○	○	○	○	○
Years of medical internship	○	○	○	○	○	○	○	○	○	○
Years of full time employment	○	○	○	○	○	○	○	○	○	○

3.

261

3. **If you have begun full time employment, please answer the following questions. (If you are still in school, skip to question #C-4) (Do not include medical internship, the Peace Corps, or Military Service other than career, as full time employment.)**

a. For each full-time employer for whom you have worked since leaving school, please mark the most appropriate category.

EMPLOYERS

	1st	2nd	3rd	4th
Private industry or business	O	O	O	O
Self employed	O	O	O	O
College or University (other than medical)	O	O	O	O
Secondary school or school system	O	O	O	O
Medical school	O	O	O	O
Federal government-civilian employee	O	O	O	O
U.S.P.H.S. - commissioned corps	O	O	O	O
Military service - active duty	O	O	O	O
State government	O	O	O	O
Local government	O	O	O	O
Non-profit hospital or clinic	O	O	O	O
Other non-profit organization	O	O	O	O
Other	O	O	O	O

b. Please indicate those full-time positions which you have held since leaving shcool. (Mark all that apply)

POSITIONS

	1st	2nd	3rd	4th
Military service (career)	O	O	O	O
Musician (performer, composer)	O	O	O	O
Nurse	O	O	O	O
Optometrist	O	O	O	O
Pharmacist	O	O	O	O
Physician	O	O	O	O
School counselor	O	O	O	O
School principal or superintendent	O	O	O	O
Scientific researcher	O	O	O	O
Social worker	O	O	O	O
Statistician	O	O	O	O
Therapist (physical, occupational, speech)	O	O	O	O
Teacher (elementary)	O	O	O	O
Teacher (secondary)	O	O	O	O
Veterinarian	O	O	O	O
Writer or journalist	O	O	O	O
Skilled trades	O	O	O	O
Other	O	O	O	O

c. Please write the specific title of your current position. (Write in shaded area only)

POSITIONS

1st 2nd 3rd 4th

Accountant or actuary .... ○ ○ ○ ○

Actor or entertainer .... ○ ○ ○ ○

Architect.................................... ○ ○ ○ ○

Artist....................................... ○ ○ ○ ○

Business (clerical) .... ○ ○ ○ ○

Business executive (management,
  administrator) .... ○ ○ ○ ○

Business owner or proprietor................. ○ ○ ○ ○

Business salesman or buyer .... ○ ○ ○ ○

Clergyman (minister, priest)................. ○ ○ ○ ○

Clergy (other religious) .... ○ ○ ○ ○

Clinical psychologist .... ○ ○ ○ ○

College teacher .... ○ ○ ○ ○

Computer programmer .... ○ ○ ○ ○

Conservationist or forester .... ○ ○ ○ ○

Dentist (inc. orthodontist).................. ○ ○ ○ ○

Dietitian or home economist .... ○ ○ ○ ○

Engineer .... ○ ○ ○ ○

Farmer or rancher .... ○ ○ ○ ○

Foreign service worker (inc. diplomat)........ ○ ○ ○ ○

Housewife................................... ○ ○ ○ ○

Interior decorator (inc. designer) .... ○ ○ ○ ○

Interpreter (translator) .... ○ ○ ○ ○

Lab technician or hygienist.................. ○ ○ ○ ○

Law enforcement officer..................... ○ ○ ○ ○

Lawyer (attorney) .... ○ ○ ○ ○

3 b continued column 2

Briefly state your principal responsibilities.

4.

d. Please indicate the income you have earned from full time employment for each year listed.

YEARS OF FULL TIME EMPLOYMENT AFTER SCHOOL

INCOME PER YEAR	1st	2nd	3rd	4th	5th	6th	7th	8th	9th	10th
Under $5,000	○	○	○	○	○	○	○	○	○	○
$5,000	○	○	○	○	○	○	○	○	○	○
6,000	○	○	○	○	○	○	○	○	○	○
7,000	○	○	○	○	○	○	○	○	○	○
8,000	○	○	○	○	○	○	○	○	○	○
9,000	○	○	○	○	○	○	○	○	○	○
10,000	○	○	○	○	○	○	○	○	○	○
11,000	○	○	○	○	○	○	○	○	○	○
12,000	○	○	○	○	○	○	○	○	○	○
13,000	○	○	○	○	○	○	○	○	○	○
14,000	○	○	○	○	○	○	○	○	○	○
15,000	○	○	○	○	○	○	○	○	○	○
16,000	○	○	○	○	○	○	○	○	○	○
17,000	○	○	○	○	○	○	○	○	○	○
18,000	○	○	○	○	○	○	○	○	○	○
19,000	○	○	○	○	○	○	○	○	○	○
20,000	○	○	○	○	○	○	○	○	○	○
22,000	○	○	○	○	○	○	○	○	○	○
24,000	○	○	○	○	○	○	○	○	○	○
26,000	○	○	○	○	○	○	○	○	○	○
28,000	○	○	○	○	○	○	○	○	○	○
30,000	○	○	○	○	○	○	○	○	○	○
Over $30,000	○	○	○	○	○	○	○	○	○	○

264

e. To what extent do you utilize your undergraduate major and graduate school area (if applicable) in your current position?

	VERY MUCH	SOME	VERY LITTLE	NONE
Undergraduate major	O	O	O	O
Graduate School area	O	O	O	O

4. The following professional activities are amongst those in which you may have engaged since leaving school. (Please make the appropriate response for each.)

NUMBER OF OCCASIONS

	1	2	3	4 or more	Not attempted	Undertaken but not achieved
Articles published	O	O	O	O	O	O
Books written	O	O	O	O	O	O
Papers delivered	O	O	O	O	O	O
Research grants	O	O	O	O	O	O
Inventions	O	O	O	O	O	O
Professional memberships (national)	O	O	O	O	O	O
Professional memberships (local)	O	O	O	O	O	O
Regional or Professional Who's Who	O	O	O	O	O	O
Post doctoral awards	O	O	O	O	O	O
Honorary awards	O	O	O	O	O	O

5.

265

SECTION D. CURRENT NON-PROFESSIONAL ACTIVITY.

The following statements describe activities that may be part of your experience since having left school. Please indicate the extent to which each activity is part of this experience.

THE ACTIVITY IS SUCH THAT
I AM INVOLVED:

	never	seldom	frequently	constantly
Read daily newspapers	O	O	O	O
Read news magazines (Time, Newsweek, etc.)	O	O	O	O
Attend fraternal organizations (Masons, K of C, etc.)	O	O	O	O
Attend civic organizations (Rotary, Kiwanis, etc.)	O	O	O	O
Participate in educational groups (P.T.A., etc.)	O	O	O	O
Hold office in local civic, fraternal, or educational organizations	O	O	O	O
Participate in service organizations (Fire Co., Red Cross, etc.)	O	O	O	O
Use a library card	O	O	O	O
Participate in local political organizations	O	O	O	O
Hold local political office	O	O	O	O
Participate as member of a club (Hobby, sports, etc.)	O	O	O	O
Read non-professional books	O	O	O	O
Serve in church or synagogue	O	O	O	O
Appointed on committees to serve the community	O	O	O	O
Read literary magazines such as the Atlantic Monthly	O	O	O	O
Attend cultural events in the community	O	O	O	O

SECTION E. YOUR OPINIONS.

1. If you chose not to undertake graduate work, what were the major influences which contributed to this decision?

5. When you compare yourself with science graduates that you have known from other institutions, how do you rate your preparation for the problems you have encountered?

266

Lack of funds.............. ○   Weak results of G. R. E. ......○
Marriage .... ○   Need to earn money ....○
Pregnancy.................. ○   Armed services commitments. ○
Poor undergraduate grades... ○   Peace Corps interests........○
Fine job immediately available ○   Other ....○

2. If you have not entered graduate school (or have left), do you intend to complete a graduate degree in the future?

No....○ Yes....○ as a: Part time student ...... ○
                                    Full time student ...... ○

3. If you have attended graduate school for at least one semester, how did you find the competition as compared to that which you encountered at college?

Less imposing than at college... ○   More imposing than at college... ○   About the same as at college... ○

4. Whether or not you have attended graduate school, what do you believe to be the quality of your undergraduate experience? (Mark all that apply)

	THE PREPARATION WAS:				PROGRAM SHOULD BE:	
	STRONG	ADEQUATE	WEAK	NO OPINION	INCREASED	DECREASED
Laboratory technique	○	○	○	○	○	○
Advanced courses	○	○	○	○	○	○
Research orientation	○	○	○	○	○	○
Breadth of knowledge	○	○	○	○	○	○
Quality of faculty	○	○	○	○	○	○
Student creativity	○	○	○	○	○	○
Development of student interest	○	○	○	○	○	○

PREPARATION PROVIDED BY:

	EXCELLENT	AVERAGE	POOR	NO RANK
My college	○	○	○	○
Other liberal arts colleges	○	○	○	○
"Prestigous" private universities	○	○	○	○
"Prestigous" state universities	○	○	○	○
Leading technical schools	○	○	○	○

6. Do you believe that your collegiate years provided:

	YES	NO
A broad experience in the humanities?	○	○
Facility with the arts of communication?	○	○

7. If you had it to do over, would you still select a liberal arts college for your bachelor's degree?

YES.......... ○

NO .... ○

6.

267

Science Faculty Questionnaire

RESEARCH CORPORATION: Study of Liberal Arts Science

SCIENCE FACULTY QUESTIONNAIRE
(covering graduating classes 1958-1967 inclusive)

College:_____

Department:_____

Within this questionnaire the term "science" refers only to
the departments of biology, chemistry, mathematics, physics,
and "pre-medicine". Enter the letter of the appropriate
response in the **box** on the left.

1. What is your current professional rank?

   A. Instructor B. Assistant Professor
   C. Associate Professor D. Professor

2. What is your age?

   A. Less than 25 D. 35-39 G. 50-54   J. 65 or over
   B. 25-29   E. 40-44 H. 55-59
   C. 30-34   F. 45-49 I. 60-64

3. How many years have you been on the faculty of this college?

   A. Less than 3 D. 10-12 G. 19-20
   B. 4-6   E. 13-15 H. 21-23
   C. 7-9   F. 16-18 I. 24 or more

4. Which of the following best express your opinion of the
   influence that membership in a liberal arts faculty plays
   in the esteem in which you and your colleagues are held
   by those professors who are now in major universities?

   A. I believe that we are held in substantial esteem.
   B. I believe that the esteem in which one is held has
      little or no relationship to the type of institution
      in which he is employed.
   C. I believe that there now exists a noticeable, though
      slight, negative attitude within the profession
      toward those of us engaged in liberal arts science.
   D. I believe that there is a very definite negative
      image associated with teaching science within a
      liberal arts institution.

5. Have your encountered instances within the past few years
   in which someone or some organization has discriminated
   between liberal arts science faculty and university
   science faculty such that they demonstrated greater
   esteem for those at the university?

   A. None
   B. Once or twice
   C. Several times
   D.. Many times
   E. Any discrimination which I have encountered has
      favored those at the liberal arts institutions.

270

◻ 6. What is your opinion regarding the involvement of able and/or advanced undergraduates in research activity?

   A. I do not believe in student research.
   B. I believe that involvement is worthwhile, but activity should be limited to project work because undergraduates are not prepared to undertake research.
   C. I believe in limited research for such students.
   D. I believe strongly in the concept of student research for virtually all majors.

◻ 7. How would you advise a graduating high school senior who has demonstrated genius or near-genius capability in your area of science?

   A. Enter a leading university.
   B. Enter a leading technical institution.
   C. Enter a leading liberal arts college.

◻ 8. In general, how would you compare the strength of undergraduate science programs offered at the leading universities with those offered at the leading liberal arts colleges, such as the one in which you teach?

   A. In general, universities offer stronger programs.
   B. The programs are relatively equal.
   C. In general, liberal arts colleges offer stronger programs.

◻ 9. In which type of institution did you receive your bachelor's degree?

   A. A university
   B. A technical institution.
   C. A liberal arts college.

◻ 10. Your undergraduate institution was:

   A. State supported.
   B. Privately endowed.

Department Chairman
Questionnaire

College:_____

Department_____

DEPARTMENT CHAIRMAN QUESTIONNAIRE
(covering graduating classes 1958-1967 inclusive)

For each numbered box enter the key letter designating your answer.

1. What is the policy of the college with respect to a student's declaration of major? (the policy in effect during the principal years of the study) A student declares his major at the end of semester

    A. 0 C. 2 E. 4 G. 6
    B. 1 D. 3 F. 5 H. 7

2. What is your opinion regarding this timing of a major declaration?

    A. The student declares too late in his college career; he often does not consider this major field until it is too late for him to obtain necessary elementary courses.
    B. The student declares too early and really has no idea of his major field of interest, resulting in too many changes.
    C. Timing is quite satisfactory.

Once students have declared majors some change directions, inevitably. Please indicate the approximate percent (of those in typical classes of the past decade) who fit each of the following categories. For questions 3 through 7 enter the letter of the appropriate percentage at the left.

3. Switched to your department.   PERCENT

4. Dropped from your department.   A. less than 5%   G. 30 -34%
                                  B. 5 - 9%   H. 35 -39%

5. Switched to other science.   C. 10-14%   I. 40 -44%
                                  D. 15-19%   J. 45 -49%

6. Switched to non-science.   E. 20-24%   K. 50 or
                                  F. 25-29%   more

7. Withdrew to university.

For questions 8 through 13 what influence did the following have upon attrition in your department?

Factor has been A. major B. minor C. none

8. Time commitment in laboratories too demanding.
9. Mathematical competency for advanced courses was lacking.
10. Unable to comprehend advance theories.
11. The glamour of space age science lost its lustre.
12. The greater excitement and personal challenge in another area won out.
13. This department lost out to a better equipped and staffed department.

For questions 14 through 18 what, in your opinion, have been
the principal factors motivating students in your department
to pursue and complete a major?

Factor has been A. major   B. minor   C. none

14. Ethnic background characteristics.
15. Desire to gain direct route to job financial security.
16. Desire to earn "prestige" from working as a scientist.
17. Continued influence of a parent or close person who works
    in science.
18. Consuming interest in the subject.
19. In the event that you have participated in a 3-2
    engineering program with one or more universities, how has
    the program fared within the past ten years? Indicate
    your choice by entering that letter in the box at the left.

    A. The response has been an increasing one.
    B. The program has decreased.
    C. No one has really promoted the program.
    D. Most students who indicate a desire to follow a 3-2
       curriculum later change their minds and remain for
       four years.
    E. The program has not proved worthwhile.

20. If you do not participate in a 3-2 program, are you in
    favor of developing such a relationship?

    A. yes   B. no

21. What is your opinion regarding the following statement?

    "Students who graduate in science have a common, innate
    characteristic: tenacity. Without this inner drive,
    they would not survive time commitments such as those
    required for laboratory sessions and individual
    projects."

    A. The statement expresses my feeling very well.
    B. I have no feeling one way or the other.
    C. I don't believe that innate tenacity has any
       correlation with a science major.

22. Do you believe that liberal arts distribution requirements
    deter students who might otherwise elect to major in your
    department?

    A. yes   B. no

23. Would you be in favor of reducing the number of distribu-
    tion requirements at your college so that majors in your
    department could be provided a greater depth and breadth
    of preparation?

    A. yes   B. no

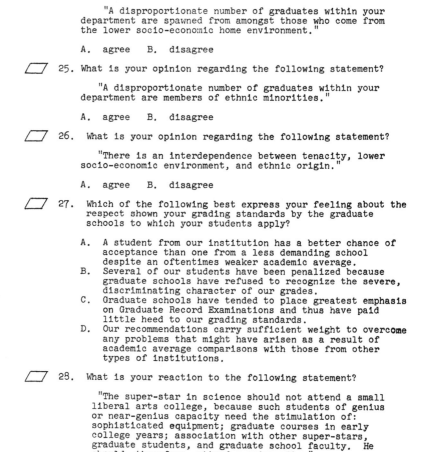

24. What is your opinion regarding the following statement?

"A disproportionate number of graduates within your department are spawned from amongst those who come from the lower socio-economic home environment."

A. agree B. disagree

25. What is your opinion regarding the following statement?

"A disproportionate number of graduates within your department are members of ethnic minorities."

A. agree B. disagree

26. What is your opinion regarding the following statement?

"There is an interdependence between tenacity, lower socio-economic environment, and ethnic origin."

A. agree B. disagree

27. Which of the following best express your feeling about the respect shown your grading standards by the graduate schools to which your students apply?

A. A student from our institution has a better chance of acceptance than one from a less demanding school despite an oftentimes weaker academic average.
B. Several of our students have been penalized because graduate schools have refused to recognize the severe, discriminating character of our grades.
C. Graduate schools have tended to place greatest emphasis on Graduate Record Examinations and thus have paid little heed to our grading standards.
D. Our recommendations carry sufficient weight to overcome any problems that might have arisen as a result of academic average comparisons with those from other types of institutions.

28. What is your reaction to the following statement?

"The super-star in science should not attend a small liberal arts college, because such students of genius or near-genius capacity need the stimulation of: sophisticated equipment; graduate courses in early college years; association with other super-stars, graduate students, and graduate school faculty. He should, therefore, attend a university."

A. agree: We are not equipped to challenge and develop the super-star.
B. disagree: We can set up an individual program with very close faculty ties such that he is developed as well as or better than one who chose to attend a university.

276

Which of the following statements best describes the principal
orientations of your department? (Make a first and second
choice, only). Enter the letter of the first choice in #29;
enter the letter of the second choice in #30.

⟋⟋ 29.   A. It is designed to prepare for graduate school.
         B. It is intended to involve students in the subject
             through research and/or project work.
         C. It is designed to provide an adequate background for
             job entry in industry or business.
⟋⟋ 30.   D. It is intended as a service to the liberal arts
             program by providing most or all students with a
             basic experience in the discipline.
         E. It is designed to supplement majors in other areas,
             such as education, with a significant minor.

What is the percent of time in the junior and senior departmental programs that have been devoted to individual project
work (as opposed to student research - see next question)?
Enter the letter representing the appropriate percent.

⟋⟋ 31. Junior year   A. 5%   C. 15%   E. 25%   G. 35%
⟋⟋ 32. Senior year   B. 10% D. 20%   F. 30%   H. 50%

If the department, or any of its members, have held research
grants which allowed the significant involvement of students,
to what extent were the students involved?

Number of students: A   B   C   7-9   E
                       0   1-3   4-6   10 or more

⟋⟋ 33. Participating during academic year 1958
⟋⟋ 34. Participating during summer 1958
⟋⟋ 35. Participating during academic year 1959
⟋⟋ 36. Participating during summer 1959
⟋⟋ 37. Participating during academic year 1960
⟋⟋ 38. Participating during summer 1960
⟋⟋ 39. Participating during academic year 1961
⟋⟋ 40. Participating during summer 1961
⟋⟋ 41. Participating during academic year 1962
⟋⟋ 42. Participating during summer 1962
⟋⟋ 43. Participating during academic year 1963
⟋⟋ 44. Participating during summer 1963
⟋⟋ 45. Participating during academic year 1964
⟋⟋ 46. Participating during summer 1964
⟋⟋ 47. Participating during academic year 1965
⟋⟋ 48. Participating during summer 1965
⟋⟋ 49. Participating during academic year 1966
⟋⟋ 50. Participating during summer 1966
⟋⟋ 51. Participating during academic year 1967
⟋⟋ 52. Participating during summer 1967

SELECTED REFERENCES

Astin, A. W., and R. J. Panos. *The Education and Vocational Development of College Students.* Washington, D.C.: American Council on Education, 1969.

Berelson, B. R. *Graduate Education in the United States.* New York: McGraw-Hill, 1960.

Bernard, Jessie. *Academic Women.* University Park, Pa.: Pennsylvania State University Press, 1964.

Chickering, A. W. *Education and Identity.* San Francisco: Jossey-Bass, 1969.

Davis, J. H. *Great Aspirations.* Chicago: Aldine, 1964.

Davis, J. H. *Undergraduate Career Decisions.* Chicago: Aldine, 1965.

Dressel, P. L., and Frances DeLisle. *Undergraduate Curriculum Trends.* Washington, D.C.: American Council on Education, 1969.

Feldman, K. A., and T. M. Newcomb. *The Impact of College on Students.* San Francisco: Jossey-Bass, 1969.

Flanagan, J. C., and W. W. Cooley, *PROJECT TALENT: One Year Follow-up Studies.* Report No. 2000. Pittsburgh: University of Pittsburgh, 1966.

Folger, J. K., Helen Astin, and A. E. Bayer. *Human Resources and Higher Education.* New York: Russell Sage Foundation, 1970.

Freedman, M. B. *The College Experience.* San Francisco: Jossey-Bass, 1967.

Harmon, L., and F. D. Boercker. *Doctorate Recipients from the United States Universities: 1958–1966.* Washington, D.C.: National Academy of Sciences, 1967.

Jencks, C., and D. Riesman. *The Academic Revolution.* New York: Doubleday, 1968.

Knapp, R. H., and H. B. Goodrich. *Origins of American Scientists.* Chicago: University of Chicago Press, 1952.

Knapp, R. H., and J. J. Greenbaum. *The Younger American Scholar: His Collegiate Origins.* Chicago: University of Chicago Press, 1953.

National Academy of Sciences. *The Invisible University: Postdoctoral Education in the United States.* Washington, D.C.: National Academy of Sciences, 1969.

National Academy of Sciences. *The Mathematical Sciences: Undergraduate Education.* Research Publication 1682. Washington, D.C.: National Academy of Sciences, 1968.

National Academy of Sciences. *Profile of Ph.D.'s in the Sciences.* OSP Publication 1293. Washington, D.C.: National Academy of Sciences, 1965.

Taylor, C. W., and F. Barron (Eds.). *Scientific Creativity: Its Recognition and Development.* New York: Wiley, 1963.

Wolfle, D. *America's Resources of Specialized Talent: A Current Appraisal and a Look Ahead.* New York: Hopper, 1954.

INDEX

Numbers in italics refer to pages on which tables appear.

Admissions office questionnaire, 2, 247–49

Articles written, by graduates, 153, *156–57,* 158

Attrition from science departments, 70–71, 72, 75–78

Awards, graduate, 170–71, 176; undergraduate (Phi Beta Kappa), 88–89, *88–89;* postdoctoral, 176–77, *178–79,* 180

Books written by graduates, 152–53, *154–55*

Breadth of knowledge, graduates' opinions on, 209, 211, *212–13,* 215, *218–19,* 223, 224–25

Choice of college, determinants in, 7–8

Civic involvement of graduates, in civic organizations, 186–87, *190–91;* in education groups, 187–88, *192–93;* in community service, *202–3,* 204

Civic organizations, membership of graduates in, 186–87, *190–91*

Club membership by graduates, 198, *198–99*

Communication, graduates' opinions on facility in art of, 234–35, *234–37*

Competitiveness in graduate school, graduates' opinions on, 235, *236–37,* 238, *238–39*

Cultural events, attendance by graduates, *202–3,* 204

Curricular requirements, as deterrents in choosing science as field, 92–94, *93, 94*

Degree aspirations, 6, 40–42; age at time of choice, 40–42, *42–45,* 43

Departmental orientations, department chairmen's opinions on, *95,* 95–96; graduate school preparation, 95, *95;* training for business or industry, *95,* 96

Department chairmen's opinions, 13–14, 92–99; on curricular requirements, 92–94, *93, 94;* on departmental orientations, *95,* 95–96; on graduate school admissions criteria, 96, *97*

Department chairmen questionnaire, 3, 273–77

Discrimination, against liberal arts science faculty, 105, *106–7,* 108, *109,* 110, 113

Education groups, membership of graduates in, 187–88, *192–93*

Esteem, professional, science faculty's perceptions of, 100–4, *102–3, 104, 106–7,* 113

Ethnic origins of graduates, 6, *18, 19,* 20–21, *20–21*

Extraprofessional activities of graduates, *see* Civic involvement; Nonprofessional organized ac-

Inventions, by graduates, 158–59, *162–63*

Liberal arts college: determinants and major influences in choice of, 46–50, *47, 48;* deterrents, 49–50; influence of guidance counseling, 49, 50; graduates' opinions on choice of, 238–39, *240–41*
Library card, use of, by graduates, 185–86, *190–91*
Literary magazines, read by graduates, 184, *186–87*
Loans, as undergraduate financial support, 55–56, *56–58,* 59
Local groups, graduate leadership in, 199, *200–1*
Local political organizations, membership of graduates in, 188–89, *192–93*

Major field of study: age at time of choice, 38–40, *38–41,* 42–43; choice of, 6, 36, 38–40
Medical school environment, as postgraduate employer, 141, *142–45*
Membership in professional organizations, by graduates, 166–67, *168–69,* 170, *170–73*
Military service, as postgraduate employer, 141, *142–45*

National Merit Scholarship Qualifying Tests, *4,* 5
Negro graduates, paucity of, 24
News magazines, read by graduates, 184, *184–85*
Newspapers, read by graduates, *182–83,* 183
Nonfraternal service groups, membership of graduates in, 194, *196–97*
Nonprofessional books, read by graduates, 184–85, *188–89*

Nonprofessional organized activities, 12, 13; fraternal organizations, 189, 194, *194–95;* nonfraternal service groups, 194, *196–97;* religious groups, 195, *196–97;* club membership, 198, *198–99;* leadership in local groups, 199, *200–1;* local political involvement, 199, *200–1,* 204; community service, *202–3,* 204; cultural events, *202–3,* 204

Occupations, types of, for graduates, 141–45, *146–49,* 150

Parental income levels, of undergraduates, 51, *52–53,* 54
Participating colleges, 245–46; affluence of, 5; "selectivity" of, 5; selection criteria, 245
Phi Beta Kappa awards, 79, 88–89, *88–89*
Political involvement of graduates: in local political organizations, 188–89, *192–93;* local office, 199, *200–1,* 204
Postgraduate employment, 11–12 duration of, 137, *138–39,* 140 nature of: primary and secondary school systems, 140, 141, *142–45;* private industry, 140, *142–45;* medical school environment, 141, *142–45;* military service, 141, *142–45;* self-employment, 141, *142–45;* state and local governments, 141, *142–45;* university or college environment, 141, *142–45* types of occupations, 141–45, *146–49,* 150
Precollege achievement, 6–7
Preparation for advanced courses, graduates' opinions on, 207, *208–9,* 211, 214, *214–15,* 216–17, *220–21*